RADIOMETRIC DATING

Edited by **Danuta Michalska Nawrocka**

Radiometric Dating
Edited by Danuta Michalska Nawrocka

Published by InTech

Janeza Trdine 9, 51000 Rijeka, Croatia

Publishing Process Manager Romana Vukelic

Technical Editor Teodora Smiljanic

Cover Designer InTech Design Team

Additional hard copies can be obtained from orders@intechopen.com

Radiometric Dating, Edited by Danuta Michalska Nawrocka
p. cm.
ISBN 978-953-51-0596-1

For EU product safety concerns:
IN TECH d.o.o., Prolaz Marije Krucifikse Kozulić 3, 51000 Rijeka, Croatia,
info@intechopen.com or visit our website at intechopen.com.

Meet the editor

Dr Danuta Michalska Nawrocka is a geologist working at the Institute of Geology at Adam Mickiewicz University in Poznań, Poland. She specialises in archaeometry, geochemistry of isotopes and geoarchaeology. Since 2000 she has cooperated with the team from the Gliwice Radiocarbon Laboratory, then with the Poznan Radiocarbon Laboratory and recently also with Circe AMS Laboratory in Caserta. Dr Michalska's doctoral thesis was devoted to the problem of dating carbonate mortars, their preparation, petrography and geochemistry. For her work on the analysis of mortars from Poland and Israel she received the Prime Minister's Prize for the distinguished doctoral theses. In recent years, she has studied the modelling of processes occurring in materials selected for dating, at different pretreatment stages.

Contents

Preface XI

Section 1 Methodology 1

Chapter 1 **Improved Sample Preparation Methodology on Lime Mortar for Reliable ^{14}C Dating 3**
Luis Angel Ortega, Maria Cruz Zuluaga, Ainhoa Alonso-Olazabal, Maite Insausti, Xabier Murelaga and Alex Ibañez

Chapter 2 **Dating of Old Lime Based Mixtures with the "Pure Lime Lumps" Technique 21**
Giovanni L.A. Pesce and Richard J. Ball

Section 2 Applications 39

Chapter 3 **Holocene Soil Chronofunctions, Luochuan, Chinese Loess Plateau 41**
Gang Liu, Wennian Xu, Qiong Zhang and Zhenyao Xia

Chapter 4 **Radiocarbon Dating in Archaeological Sites Chronology 57**
Danuta Michalska Nawrocka, Małgorzata Szczepaniak and Andrzej Krzyszowski

Section 3 Luminescence and Radiocarbon Measurements 73

Chapter 5 **Geochronology of Soils and Landforms in Cultural Landscapes on Aeolian Sandy Substrates, Based on Radiocarbon and Optically Stimulated Luminescence Dating (Weert, SE-Netherlands) 75**
J.M. van Mourik, A.C. Seijmonsbergen and B. Jansen

Chapter 6 **Luminescence Dating as Comparative Data to Radiocarbon Age Estimation of Morasko Spherical Depressions 115**
Wojciech T.J. Stankowski and Andrzej Bluszcz

Preface

This book explores the diversity of topics related to radiometric dating, with particular emphasis on the method of radiocarbon dating and a cross-check of its results with luminescence measurements.

Chemistry and the global carbon cycle is an essential element of life on Earth, because carbon is a component of every living cell, plants and animals. The radioactive isotope ^{14}C was found useful for dating, for which in 1960 Willard Frank Libby received the Nobel Prize in Chemistry. Since then, the radiocarbon method has been developing rapidly and dynamically, enabling the dating of smaller and smaller samples with high precision. The radiocarbon method is widely applied in various fields of science such as geology, archaeology, paleobotany, paleoclimatology and environmental studies.

Starting from the chapter on *Methodology* the book includes e.g. the description of the problem of preparation of samples for ^{14}C measurement, a wide application of the radiocarbon method and a comparison of results obtained by various methods, including the radiocarbon method, the method of OSL, TL and palynology. The issue of radiocarbon dating of mortars and plasters is thoroughly discussed in the book. Taking into consideration the fact that the mortar, as a material closely related to the construction of the building, may reflect the actual time of erecting the structure, the careful selection of samples and working out the rules of their preparation, which makes it possible to avoid the so-called dead carbon effect or the rejuvenating effect related e.g. to the recrystallisation of carbonates, are of great importance. Chapter Two, *Applications*, and Three, *Luminescence and Radocarbon Measurements*, provide examples of the application of the radiocarbon method in the study of archaeological sites, from the analysis of soils, loesses, to the study of organic deposits filling the depressions in the Morasko Meteorite Nature Reserve.

A wide range of different studies shows a great potential of the radiocarbon method, and the presented papers reflect the interdisciplinary research and combine the conclusions from radiometric dating, palynology and relative chronology of archaeological sites.

I would like to thank all the authors for their involvement in the process of publishing this book. Without their participation and patience during the long process of compiling this volume, its publication would not be possible. All colleagues who have agreed to review the received papers are thanked for their help.

I would like to express my special gratitude towards the Editor Book Manager - Ms. Romana Vukelic, for coordinating all activities associated with the process of compiling the book and for the continuous and very friendly contact with the authors. We are much obliged to all those who contributed to the book: reviewers, language editors, and all those impossible to mention by name.

And last but not least, very special and heartfelt thanks to our families who provided us with strength and willingness to work.

Dr. Danuta Michalska Nawrocka
Institute of Geology
Adam Mickiewicz University
Poland

Section 1

Methodology

1

Improved Sample Preparation Methodology on Lime Mortar for Reliable ^{14}C Dating

Luis Angel Ortega[1], Maria Cruz Zuluaga[1], Ainhoa Alonso-Olazabal[1], Maite Insausti[2], Xabier Murelaga[3] and Alex Ibañez[4,5]

[1]Mineralogy and Petrology Department,
[2]Inorganic Chemistry Department,
[3]Stratigraphy and Palaeontology Department, Science and Technology School,
[4]Social Sciences Department, School of Education,
The University of the Basque Country,
[5]Historical Archaeology Department, Aranzadi Society of Science,
Spain

1. Introduction

Dating ancient buildings and establishing construction phases are important issues for archaeologists and cultural heritage researchers alike. When using radiometric dating to this end, the fundamental requirement consists in acquiring suitable datable material that records the age of the studied object.

Plaster and mortar are composite building materials comprising a mixture of binder and aggregates. Binders in archaeological buildings consist of lime and gypsum, whereas aggregates contain inorganic and organic materials (Sickels, 1981). The most common organic component is charcoal, most likely corresponding to residues of the burning process, especially when a continuous wood-fired kiln was used.

Up until now, building lime, sand, pottery, and organic materials from mortars have been used in dating (Folk & Valastro, 1976; Tubbs & Kinder, 1990; Heinemeier et al., 1997a; Schmid, 2001; Goedicke, 2003; Hale et al., 2003; Benea et al., 2007; Lindroos et al., 2007; Wintle, 2008). Organic materials are widely used to date mortar and plaster (Berger, 1992; Van Strydonck et al., 1992; Frumkin et al., 2003; Rech et al., 2003; Rech, 2004; Wyrwa et al., 2009; Al-Bashaireh & Hodgins, 2011) where no other easier datable material is present, such as written inscriptions, coins, and/or historical records (see Heinemeier et al., 1997b; Hale et al., 2003; Heinemeier et al., 2010). Tubbs and Kinder (1990) reported the unreliability of dating mortar based on organic inclusions because of the old wood problem. Recently, Heinemeier et al. (2010) have presented extensive examples of the same problem and also reached the same conclusion.

The lime mortar binder represents an often-used tool to assess the chronology of the different construction phases of buildings by means of radiocarbon dating. The principle of radiocarbon dating is that binder carbonates absorb carbon dioxide from the atmosphere, thus making mortars potentially suitable for ^{14}C dating. The basis of mortar dating can be summarized as follows: limestone (mainly $CaCO_3$) is burned to lime (CaO) and this calcium oxide is then slaked with water to form portlandite ($Ca(OH)_2$). The calcium hydroxide is mixed with sand and water to make the building mortar and, during hardening, the lime mortar absorbs carbon dioxide (CO_2) from the atmosphere to produce calcium carbonate ($CaCO_3$). During the hardening process, the actual ^{14}C concentration of CO_2 in air is fixed to the binder carbonate.

The method of radiocarbon dating has been applied since the 1960s (Labeyrie & Delibrias, 1964; Stuiver & Smith, 1965; Baxter & Walton, 1970; Van Strydonck et al., 1983; Van Strydonck et al., 1986; Ambers, 1987; Van Strydonck et al., 1992; Heinemeier et al., 1997a; Hiekkanen, 1998; Hale et al., 2003; Nawrocka et al., 2005; Lindroos et al., 2007; Nawrocka et al., 2007; Nawrocka et al., 2009; Heinemeier et al., 2010; Marzaioli et al., 2011), but frequently the isotopic age and the expected historic age differ, with the former providing older dates.

In dating building mortars, problems to obtain the correct age are common, often related to an adequate selection of binder carbonate. Unfortunately, most lime samples contain carbons of different provenances: incompletely burned limestone fragments, charcoal particles from wood-fired kilns, and aggregates. Aggregates are mainly natural sands. Depending on the geological and geographical location, it may also contain carbonate minerals, carbonate rock fragments, and calcareous fossils. So, the aggregates primarily constitute an additional source of contamination in carbon. Other potential problem in mortar dating is related to the hardening process. In view of the radio-chronological results some authors suggest that mortar lying on the inside of walls or behind stone facing can take years and decades (Van Strydonck & Dupas, 1991) and even centuries for voluminous structures (Sonninen et al., 1989) before the whole construction has carbonated. Thus yielding a date that is too young for the building as a whole. However, based on experimental works the hardening process of a mortar is relatively rapid compared with the half-life of ^{14}C (Pachiaudi et al., 1986; Lanas et al., 2005; Stefanidou & Papayianni, 2005; Kosednar-Legenstein et al., 2008; Ball et al., 2011). According to these researches the hardening is completed within weeks or months. Since the binder phase is constituted of small calcite grains (e.g. Stefanidou & Papayianni, 2005; Zamba et al., 2007; Ball et al., 2011; Marzaioli et al., 2011), an accurate selection of adequate lime grain-size fraction allows to date this rapid carbonation reaction.

Therefore, in order to eliminate these error sources, sample preparation procedures have been improved upon since the beginning of radiometric method dating. Most of these preparations consist of mechanical pre-treatment and chemical treatment (Sonninen & Jungner, 2001; Lindroos et al., 2007; Nawrocka et al., 2007; Goslar et al., 2009; Nawrocka et al., 2009; Heinemeier et al., 2010).

In mechanical separation, the mortar samples are gently broken and then sieved (different sized meshes are used). Van Strydonck et al. (1992) used a 250 μm sieve, whereas others (Heinemeier et al., 2010 and references therein) utilized an increasingly fine mesh ranging between 20–500 μm. Following mechanical separation, the mineral composition is analysed by petrographic microscopy supplemented with cathodoluminescence in order to identify binder contamination from aggregates and unburnt limestone (Heinemeier et al., 2010 and references therein). In chemical separation, binder carbonate is gradually dissolved by pouring 85% phosphoric acid over the powder mortar under vacuum (Heinemeier et al., 2010), under the assumption that mortar binder carbonate dissolves much more easily than limestone. However, significant amounts of detrital carbonate and other carbon sources are also dissolved, therefore precluding the determination of the correct age.

In addition, mechanical methods involving manual disaggregation have been used as an alternative to the wet chemical method for mortars with calcareous aggregates (Cimitan et al., 1991; Casadio et al., 2005; Nawrocka et al., 2005; Ortega et al., 2008). Mechanical binder separation is recommended when carbonate contamination has been observed during petrographic analysis.

An effective separation method of very pure binder fraction to assess radiocarbon dating is proposed in the present contribution. The elimination of contaminant error sources and the suitable selection of mortar samples enable the reliable dating. Petrographic analysis under polarizing light microscopy is used to identify different mineralogical phases of mortar samples, aggregate nature, and the limestone fragment remains. It also provides the identification of possible features of mortar degradation (organic and inorganic) and, therefore, allows the selection of a suitable sample for radiocarbon dating. To obtain datable binder, thin-section assessment is essential and offers one of the most effective methods for mortar sampling. A more detailed microscopic analysis can be provided by Scanning Electron Microscopy, which allows contaminant error sources of the mortar sample on a small scale to be identified.

Once the aggregate nature and the occurrence of other carbonate materials are established, the binder carbonates are separated by a combined mechanical and physical procedure (Ortega et al., 2008). This method removes the carbonate fraction, lime lumps and the charcoal particles. The extraction procedure allows to obtain binder reliable for dating without using partial acid digestion and several radiocarbon measurements of complex interpretation. In order to test the effectiveness of mechanical separation and to verify the purity of the binder, Scanning Electron Microscope (SEM), X-ray diffraction (XRD) analyses and thermogravimetric analysis (TGA) were performed. To test the developed procedure, historic lime mortars from the parish church of Santa Maria la Real (Zarautz, northern Spain) have been ^{14}C dated.

2. Materials and methods

The mortars were thoroughly examined in the laboratory using a stereo-zoom microscope and carefully disaggregated to avoid breaking the existing aggregates.

Thin sections and polished sections of the mortars were prepared in an impregnation unit under vacuum with an epoxy resin. These were polished with 15 µm Al_2O_3 abrasive and the final lapping was performed with diamond pastes (15–1 µm). Thin-sections were analysed by light polarized microscopy using a petrographic polarizing Olympus BH2 microscope equipped with an Olympus DP-10 digital camera. Scanning electron microscopy observations were performed with a JEOL JSM-7000F Schottky-type field emission scanning electron microscope operating with an Oxford Pentafet photon energy instruments Link Isis X-ray (EDX) microanalysis system. Samples were carbon-coated to eliminate charging effects. The cathodoluminescence (CL) study was performed using a

Technosyn Cold Cathode Luminescence system, model 8200 Mk II, mounted on an Olympus trinocular research microscope with a maximum magnification capability of 400 x, using universal stage objectives. Standard operating conditions included an accelerating potential of 12 kV and 0.5–0.6 mA beam current with a beam diameter of approximately 5 mm. X-ray diffraction (XRD) was carried out using monochromatic Cu-$k_{\alpha 1}$ X-radiation at 40 kV and 20 mA, speed of 0.05 °/s and 2θ ranging from 3 to 74 from a Philips PW1710 diffractometer.

Thermogravimetric and differential thermal analysis were also performed in a TA SDT 2960 TG-DSC simultaneous instrument. Pt crucibles containing 5-7 mg of sample were heated at 2°C/min from room temperature to 1000 °C under dry oxidizing atmosphere. In order to verify the effectiveness of the separation method, several fractions during extraction process have been analyzed by above described techniques.

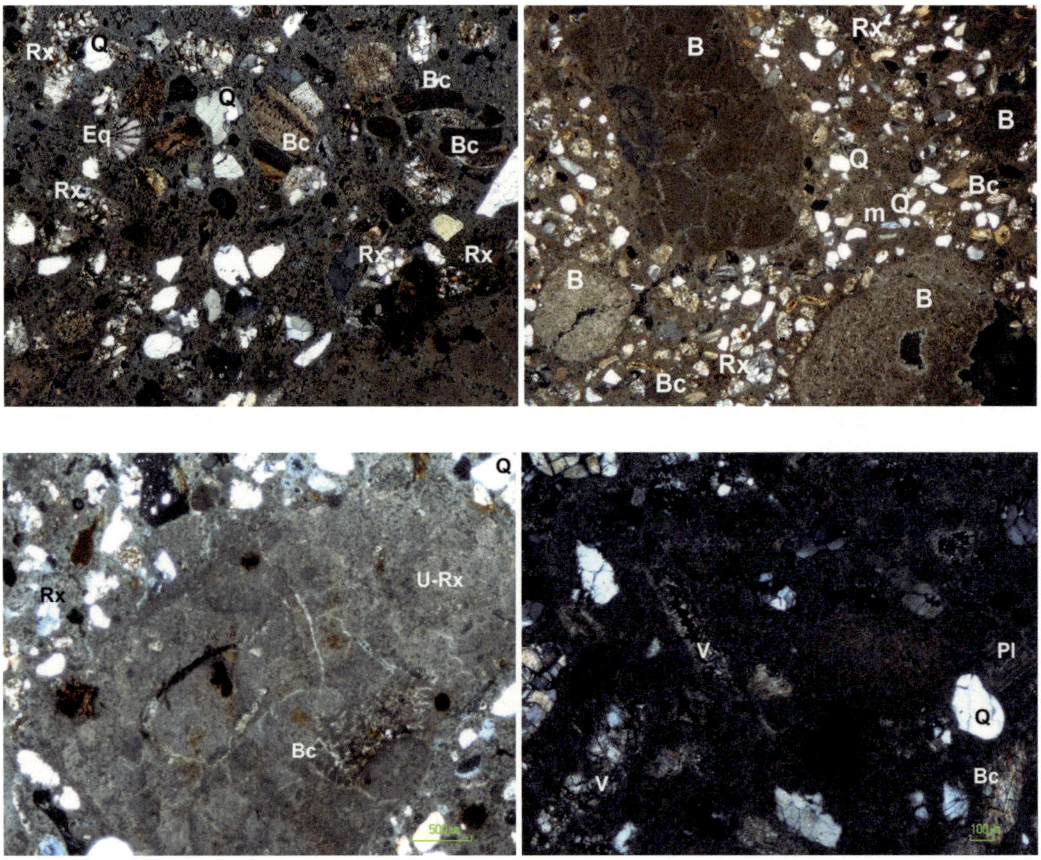

Fig. 1. Photomicrographs of historic lime-mortars. **A.** Binder with rounded quartz (Q), sandstones (Rx) and biogenic carbonate fragments (Bc: bioclasts, Eq: echinoderm). **B.** Carbonate *"pure binder"* globules (B) with fractures. Rx: rock fragments, m: muscovite, Q: quartz, Bc: bioclasts . **C.** Fossils inside of an unburned limestone fragment (U-Rx). Bc: bioclasts, Rx: rock fragments, Q: quartz. **D.** Void (V) filled with secondary carbonate. Q: quartz, Bc: bioclasts, Pl: plagioclase.

Before sample preparation for dating, the mortar samples were subjected to petrographic analyses in order to determine binder and aggregate types. Thin-section analysis shows that beach sand aggregates were used as a mortar mixture in all samples. Rounded detrital quartz fine grains, carbonate shell fragments, mainly briozooa and ostracods (Fig. 1A) are common. Inorganic detrital carbonate fragments (micritic and sparitic limestone) has been hardly observed (Ortega et al., 2008). Pure carbonate globules, visible at a macroscopic and microscopic scale, are common (Fig. 1B). In previous works, these globules have been explained as not complete homogeneous mixture of lime binder and aggregates featured by the presence of retraction fractures (Nawrocka et al., 2005). However in this study, most of the globules constitute an incomplete burned limestone as suggest the occurrence of fossil fragments within these globules (Fig. 1C). The petrographic study also reveals the occurrence of voids and fractures in the mortars refilled with secondary carbonate (Fig. 1D).

Once established the mortar nature and determinate the grade of alteration cathodoluminescence analyses was performed over mortar thin section. This method is very useful complement to petrographic microscopy and is especially sensitive when it comes to distinguishing between carbonate phases with slightly different crystallinity and trace element chemistry (Marshall, 1988; Machel, 2000; Pagel, 2000). Natural calcites and dolomites in sedimentary rocks show luminescence in different colours and intensities revealing complex history of cements and crystal growth history (Marshall, 1988). The causes of the luminescence are defects in the crystal structure or impurities hosted in the crystal lattice. The cathodoluminescence of carbonates is mostly due to the presence of transitional elements and, in particular, to Mn^{2+} as a main activator ion (orange luminescence); Fe^{2+} is believed to be the most important quencher ion (Marshall, 1988; Corazza et al., 2001). The most important factor determining the colour and intensity of the luminescence is the Mn^{2+}/Fe^{2+} ratio in the calcite (Habermann et al., 2000). The chemical activity of these ions is controlled by the decreasing pH in crystallizing calcite. So, is possible to assess how many carbonate contaminants are in the mortars by performing image analyses, but the results are at best only semi-quantitative due to many problems (Lindroos et al., 2007).

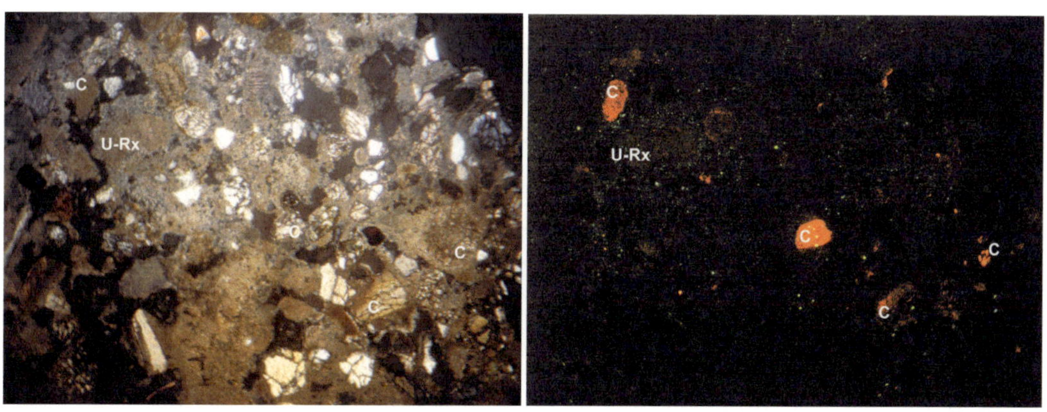

Fig. 2. Photomicrographs of historic lime-mortars. **A.** Unburned limestone (U-Rx) remains and sparite limestone detrital grains (C). **B.** Cathodoluminescence image.

The binder phase is typically a spongy network of microscopic crystal of calcite enclosing the aggregate particles as reveals the SEM study. During mortar cementation, the expelled water generates porosity and at the same time transports dissolved atmospheric CO_2 to crystallize $CaCO_3$. In cathodoluminescence, the binder is usually dark brown or nearly black (Fig. 2B) revealing a rapid hardening process in contact with atmospheric oxygen. This very oxidizing conditions leads to the precipitation of both Mn and Fe as oxides and hydroxides, and therefore preclude the diffusion of activator elements in the calcite lattice.

Grains with well-developed bright yellowish to red luminescence reveal the occurrence of carbonates with diagenetic modification and correspond to fragments of different inherited carbonate types included in the aggregates. The dully-luminescent grains correspond to residues of unburned limestone (Fig. 2).

3. Extraction procedure

The protocol established to obtain pure datable binder is based on a modification of the procedure described by Sonninen and Jungner (2001) and largely used in the radiocarbon dating of mortar (Nawrocka et al., 2005; Lindroos et al., 2007; Nawrocka et al., 2007; Goslar et al., 2009; Nawrocka et al., 2009; Heinemeier et al., 2010). The mortar samples were gently crumbled up manually or with a mortar and, based on the assumption that binder carbonates are characterized by an easily breakable aggregation structure, they were disaggregated by means of an ultrasonic bath.

This method is based on the fact that particle size is related to mechanical or chemical origin. Particles of mechanical origin always have a grain size of over 1 µm, whereas chemical reactions produce colloids that flocculate and regrow, generating finer particles. This process has been demonstrated for subaerial media (Wilson & Spengler, 1996) and for aqueous media (Davis & Kent, 1990; Salama & Ian, 2000). Therefore, this work has optimized a procedure to obtain particles of under 300 nm, which ensures that all the carbonate separated has been generated by slaked lime carbonation, and consequently that the carbon measured corresponds to atmospheric carbon.

The methodology is a variation on the particle-fractionation techniques routinely used in soil mineralogy studies. Sedimentation and centrifugation allow the routine separation of particles of < 2 µm (Laird & Dowdy, 1994; Soukup et al., 2008). Particle fractionation is based on the differential settling of the particles in a liquid so that centrifugation increases the rate of sedimentation, accelerating the process. The relation between the sedimentation rate and particle size is given in Stoke's Law (Stokes, 1851):

$$V = g(s_p - s_l)D^2/1.8h$$

where V is the particle velocity (cm/s) in the liquid of density (s_l); g is gravity acceleration (9.8 cm/s^2); s_p is the particle density (g/cm^3); D is the equivalent spherical diameter of the particle (cm); and h is the viscosity of the liquid (Pa.s).

The main goal of this separation procedure is to avoid (or minimize) the aggregation of small particles and to fractionate particles of under 1 µm. To explain this extraction method, the procedure has been designed in different particle-fractionation steps. (i) The manually crumbled mortar is placed in an ultrasonic bath of ultrapure water for 10 minutes to promote further crumbling. Afterwards, the suspended fraction is extracted. (ii) This fraction is purified through centrifugation at 3000 rpm for 10 minutes using a Kubota 3000 centrifuge (Kubota Corporation, Tokyo, Japan). (iii) Then, the topmost 50 ml by volume is collected, corresponding to the under 20 µm fraction. In order to assess the particle size obtained, a Mastersizer 2000 particle analyser (Malvern Instruments Ltd, Malvern, UK) was used to measure grain size. Figure 3A gives a particle-distribution histogram with two modes, one for fine particle size (0.2 µm) and the other for coarse (≈ 1.5 µm), with a tail reaching 20 µm. (iv) In order to improve the grain-size separation of this < 20 µm fraction, it is resuspended in an aqueous medium at a pH ≈ 8 and then placed in the ultrasonic bath to promote crumbling. The pH ≈ 8 aqueous medium favours optimal scattering of small crystals (Warkentin & Maeda, 1980). This suspension is centrifuged for one minute at 1000 rpm, after which the topmost 15 ml are collected, which corresponds to a grain-size fraction of less than 1 µm and a mode of 200 nm (Fig. 3B and Fig. 5). The XRD tests are performed to ensure that the extraction method was effective for lime-binder refining. The final fractions thus obtained were composed only by calcite (Fig. 4A) whereas previous fractions from the step (iii) content some residual aggregates of clay minerals and quartz nature (Fig. 4B). This procedure is repeated as necessary to obtain a sufficient amount for AMS analysis. In this study, this process had to be repeated five to eight times to obtain 40 to 80 mg.

This method separates the carbonate binder particles (formed through the reaction of the lime with atmospheric carbon) from the detrital and/or fragmented particles corresponding to aggregates and unburned limestone. Therefore, this extraction procedure eliminates any supply of dead or non-atmospheric carbon. The separation of the carbonate binder has been confirmed by SEM microscopy observations. The figure 5A shows planar carbonate grains of the binder with a grain size of approximately 0.2 µm in the useful fraction after step (iv). The coarse fraction discarded can be observed in figure 5B. This fraction commonly contains particles with a grain size of over 1 µm and habits suggesting an inherited origin of the carbonate.

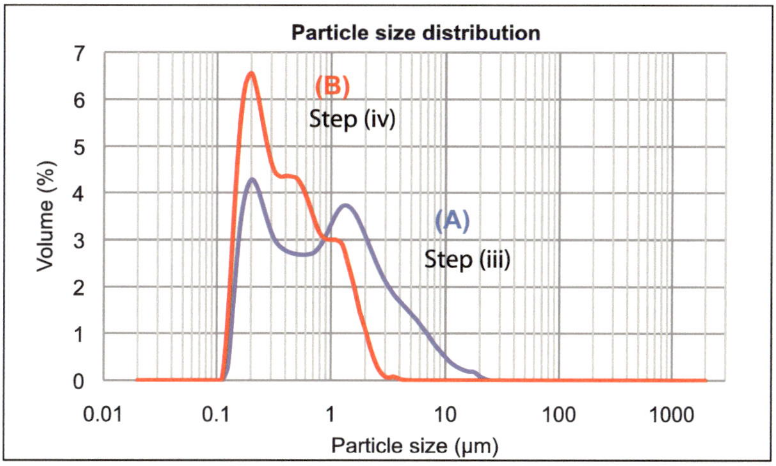

Fig. 3. Grain-size histograms of particles extracted from roman mortar (sample SNR-579). **A.** The distribution after extraction step (iii) exhibits bimodality with a large number of particles in a mode at small sizes (0.2 µm) and coarse size mode (≈1.5 µm) and with a tail reaching to 20 µm. **B.** The distribution after extraction step (iv) shows a mode at 0.2 µm and to shoulder at 0.5µm and 1µm with a little tail reaching to 4 µm.

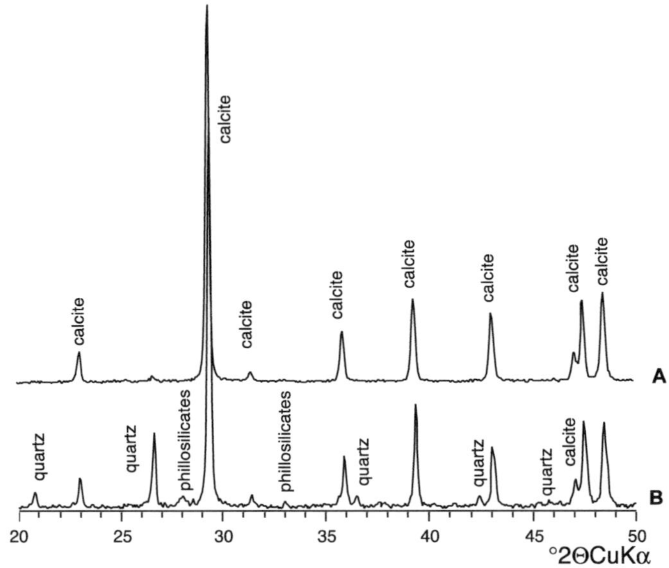

Fig. 4. XRD patterns of binder mortar. A. After extraction step (iv). B. The "coarse fraction" after extraction steep (iii).

Fig. 5. SEM photomicrographs of extracted carbonate from SNR-108 mortar sample. **A.** Fine fraction with average particle size ca 0.2 nm. **B.** Larger fraction with detrital or mechanically fragmented inherited carbonates.

Moreover, thermogravimetric analysis and differential thermal analysis (DT-TGA) has been carried out to obtain a complementary knowledge about the binder composition and purity. The figure 6 illustrates representative thermograms obtained for studied samples. The thermograms of all samples are typical of aerial lime mortars of carbonate nature with a typical weight loss at temperatures ranging from 600 to 750 °C (Bakolas et al., 1998; Marques et al., 2006; Adriano et al., 2009).

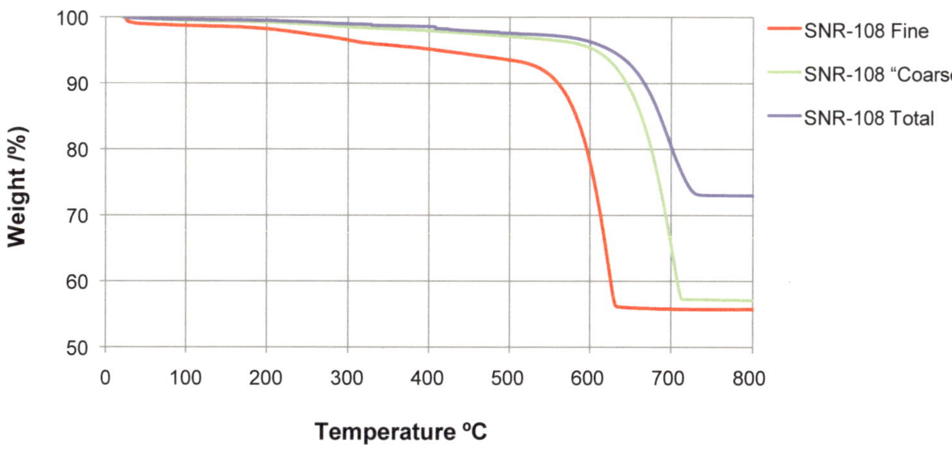

Fig. 6. Thermal analysis curves of mortar and different binder fractions of SNR-108 sample.

The finer grain size fraction corresponding to pure binder reacts at lower temperatures than the coarse size fraction and bulk mortar, showing the shifting of weigth loss temperature from ca 700 ºC to 600 ºC as purity increases and grain size decreases. The continuous weight loss observed between 200ºC and 550 ºC in the lime fine fraction could be attributed to $Ca(OH)_2$ dehydration (Paama et al., 1998).

The shifting of temperatures reveals that this fine fraction corresponds to the most reactive fractions of the sequential solutions performed by other researchers (Sonninen & Jungner, 2001; Lindroos et al., 2007; Nawrocka et al., 2009; Heinemeier et al., 2010). The main advantage of this extraction method consists in the removal of any inherited carbonate contamination.

To test the effectiveness of the developed extraction method, five mortar samples were selected for radiometric dating corresponding to different periods ranging from Roman (1st to 4th centuries AD) to Medieval (9th to 14th centuries).

4. Results and discussion

Santa Maria la Real is located within and surrounding the parish church of the Zarautz town (northern Spain). The town is located on the Gulf of Biscay, in front of a sandy beach two kilometres long. Excavations of the interior of the church of Santa Maria la Real have identified successive periods of occupation since the Iron Age (Ibáñez Etxeberria, 2003; Ibáñez Etxeberria & Moraza, 2005; Ibañez Etxeberria, 2009) (Fig. 7). Remains from the Iron Age and a long-term stable Roman establishment (between the 1st to 5th centuries AD) were located in the room structure uncovered. Within the Roman zone, several room units have been discovered. A new reoccupation of the space by different human groups took place starting in the late 9th century or early 10th century.

Radiocarbon dating of the lime mortars has revealed the age of various construction phases at the temple of Santa Maria la Real (Table 1). The earliest age determined ranges between 90 - 210 AD (with a 95.4% probability) and corresponds to construction phases at the end of the 1st or 2nd centuries of the present era (Fig. 9). This age is in agreement with the relative timeline indicated by the abundance of common Early Empire Roman pottery and with the remains of Hispanic terra sigillata pottery found at the site (Ceberio, 2009; Cepeda Ocampo, 2009; Esteban Delgado et al., in press). The oldest remains of Hispanic terra sigillata pottery correspond to pottery produced at *Tritium magallum* from 40 AD to 60 AD, equivalent to the earliest age of foundations for the Roman structures. In fact, the fill of a foundation trench for the main structure preserved from the Roman period contained fragments of Hispanic terra sigillata pottery (Cepeda Ocampo, 2009) and of common hand-made pottery (Ceberio, 2009), as well as a carbon fragment in the mortar. This carbon fragment was dated by ^{14}C at 1930±40 (Ua-20919), corresponding to a calibrated age of 40 BC to 140 AD with a 95.4% probability (Ibañez Etxeberria & Sarasola Etxegoien, 2009). These ages are perfectly consistent since in ^{14}C dating calibration, one has to take into account the problem of old wood (Bowman, 1990). Wood or carbon fragments in a mortar are always older than the mortar itself. The difference can be significant if the carbon analysed derives from the internal part of the tree trunk used as fuel to produce the mortar, or even more so if the wood comes from trees cut long before.

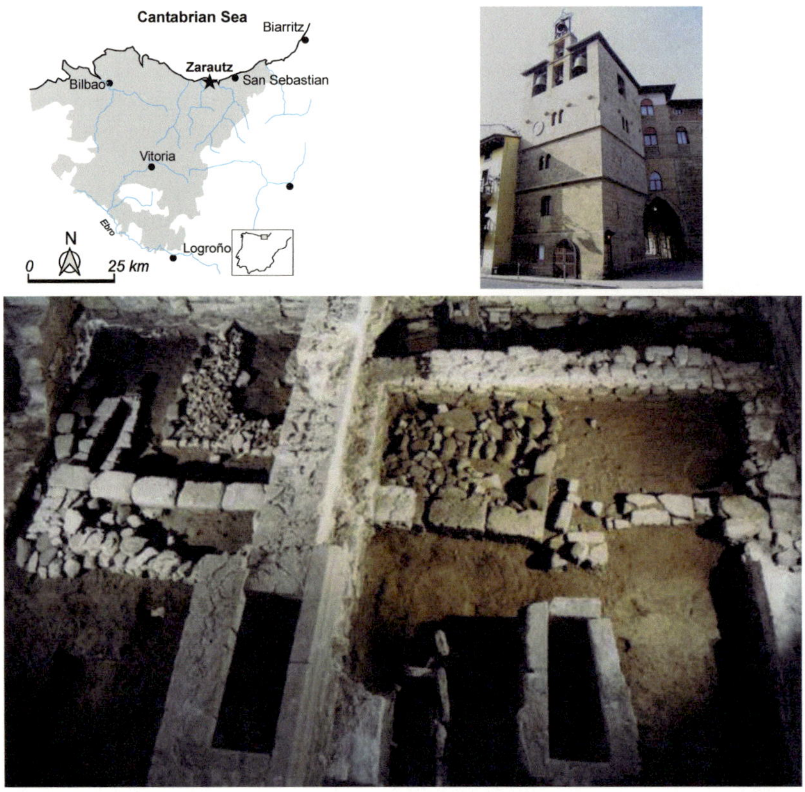

Fig. 7. Geographical location of the church of Santa Maria la Real and one temple view.

The evolution of the site is attested through the layout of four religious temples and an associated necropolis (Fig. 8). Overall, this confirms the foundation and the uninterrupted development of the present-day community of Zarautz from the early and late Medieval periods until the existence of the present temple.

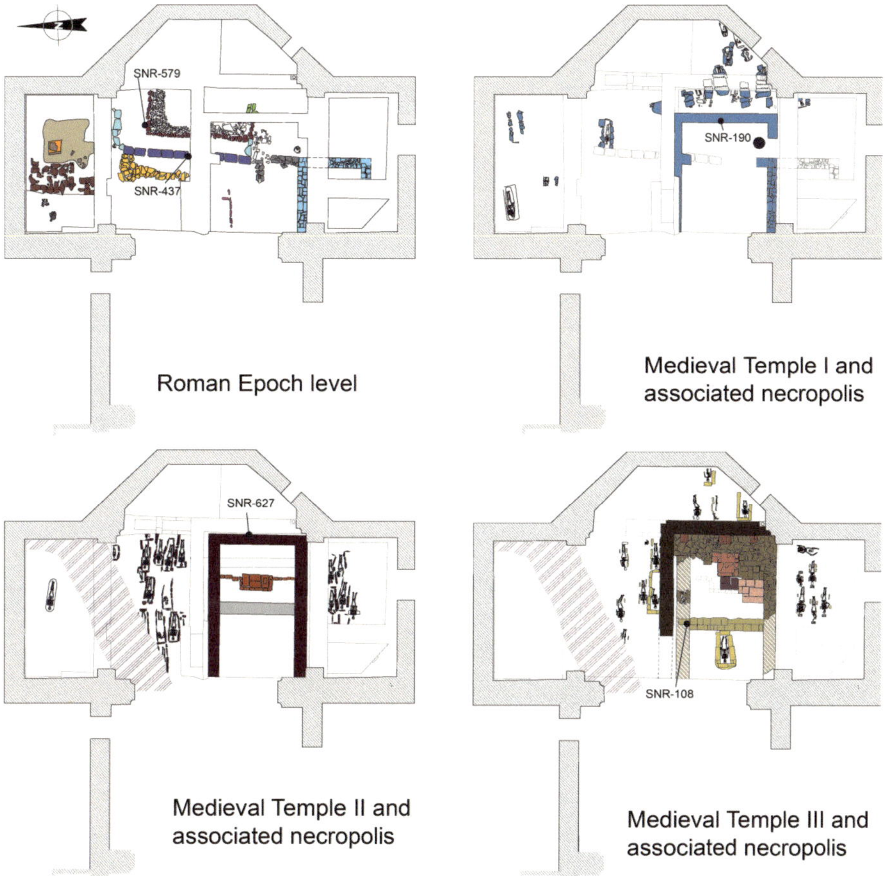

Fig. 8. Scheme of the studied temples and the location of mortar samples.

Lab Code	Sample	Conventional Age	$\delta^{13}C$	Calibrated Age		Archaeological Age
				68.20%	95.40%	
Beta-300900	SNR-579	1860 +/- 30 BP	-9.4 ‰	90 - 214 AD	80 - 231 AD	Roman
Beta-300899	SNR-437	1870 +/- 30 BP	-10.5 ‰	92 - 210 AD	73 - 227 AD	Roman
Beta-300898	SNR-190	1220 +/- 30 BP	-17.3 ‰	729 - 870 AD	692 - 888 AD	Temple I
Beta-300901	SNR-627	1100 +/- 30 BP	-17.4 ‰	897 - 985 AD	887 - 1014 AD	Temple II
Beta-300897	SNR-108	1060 +/- 30 BP	-11.8 ‰	905 - 1019 AD	896 (16.2%) 920 AD 950 (79.2%) 1024 AD	Temple III

Table 1. AMS ^{14}C dates for mortar samples of Santa Maria la Real obtained with OxCal v 4.1 (Bronk Ramsey, 2009) and IntCal09 atmospheric data (Reimer et al., 2009).

This date belongs to the earliest years of Roman occupation on the northern coast of Spain, related with the creation of coastal sites for maritime trade, which took place especially starting in the second half of the 1st century AD (Fernández Ochoa & Morillo Cerdán, 1994; Esteban Delgado, 2004). In contrast, despite the persistence of this settlement for various centuries during the Roman period, as shown by the finding of Late Empire coins and common pottery (3rd–5th centuries) (Cepeda Ocampo, 2009; Esteban Delgado et al., in press), no significant later structures have been dated. The main reason is that, at this site, the later buildings documented are of poorer construction, with walls mortared with mud (Ibañez Etxeberria, 2009).

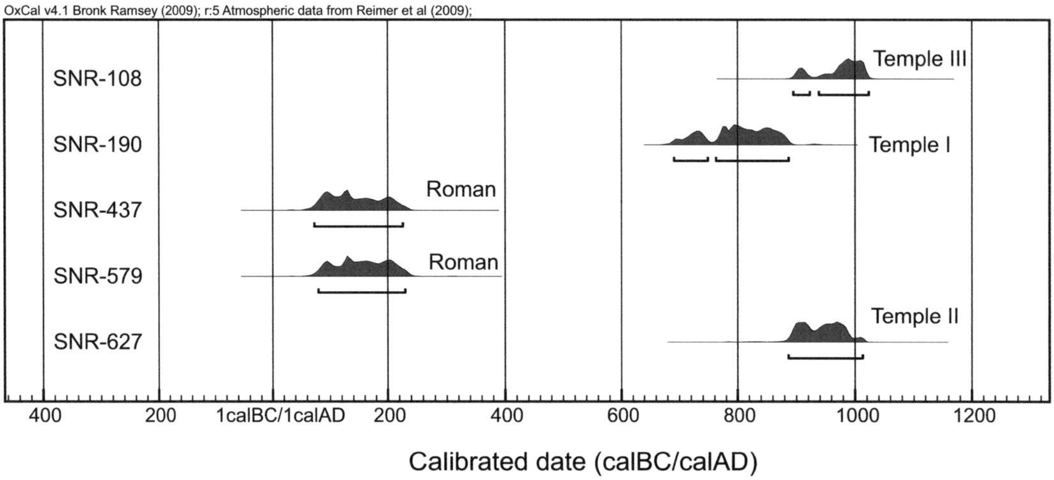

Fig. 9. Calibration of ^{14}C dates of the Santa Maria la Real mortar samples, obtained with OxCal v 4.1 (Bronk Ramsey, 2009) and IntCal09 atmospheric data (Reimer et al., 2009).

For the Medieval period, the dates obtained correspond to the construction of the perimeter walls for various buildings of the parish church. These buildings are built one atop the other on the same religious site (Fig. 8) due to greater need for space as a result of the population growth of Zarautz in the Early Medieval period. The oldest age, consistent with the archaeological findings, is provided by the mortar in wall SNR-190 (Temple I), with a probable age of 772 AD to 870 AD (68.2% confidence interval) (Fig. 9). This dating is close to that established by archaeological analyses for an arch of the Temple I, and therefore is not likely a spurious aged dating. Slightly older ages were obtained in skeletons from burials associated with Temple I. Thus, skeleton sample Ua-16897 provides an imprecise age of 558 AD to 870 AD (95% confidence interval), and probably prior to 770 AD (68.2%) (Ibañez Etxeberria & Sarasola Etxegoien, 2009).

The other two ages belong to the perimeter walls of Temples II and III. The age for Temple II must be prior to 985 AD (68.2%), whereas Temple III provides an age of 970 AD to 1020 AD (95%). These two construction phases give ages very close to each other, which is consistent with the population surge at the end of the turn of the millennium and the subsequent need for a larger temple.

The ages determined are coherent not only with the archaeological information, but also with previous radiometric dating (Fig. 10). The age for Temple II is very close to those determined for the burials in stone-slab graves in the necropolis, with a dating of 890 AD to 1040 AD (68% confidence interval), whereas individuals associated with the Temple III necropolis, with walled graves, give ages of 990 AD to 1160 AD (Ibáñez Etxeberria, 2003).

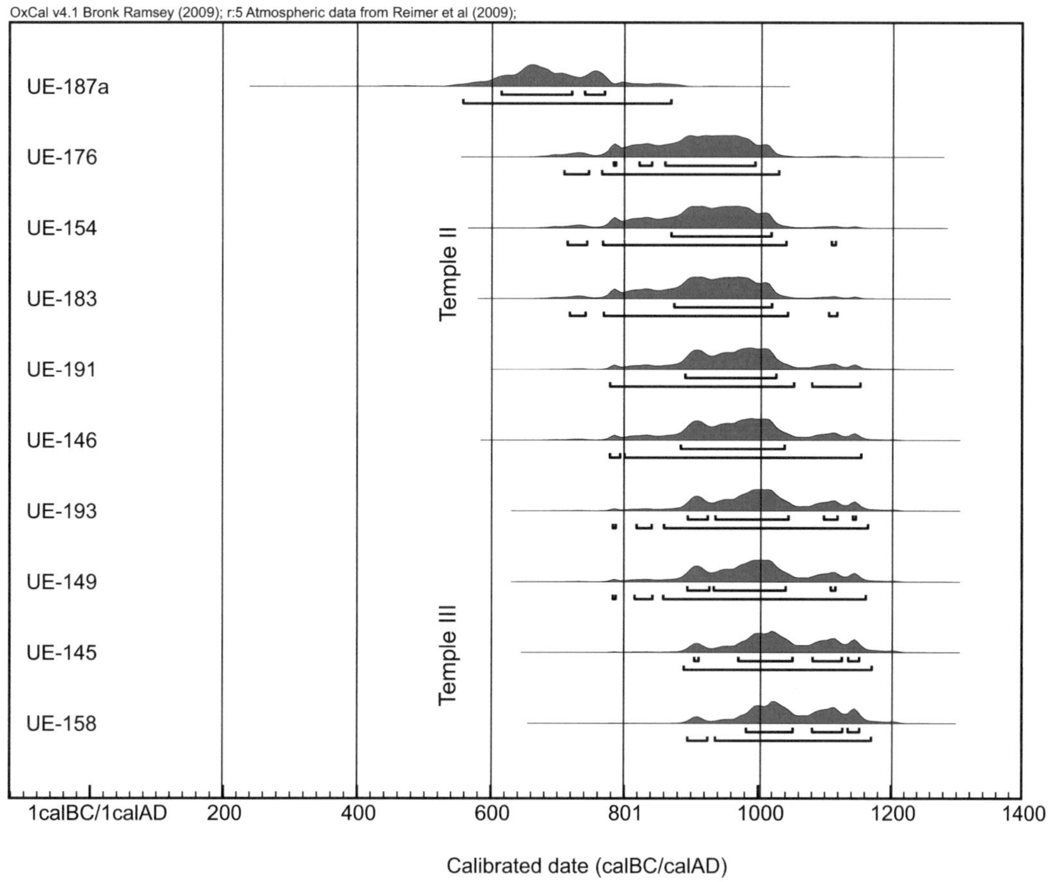

Fig. 10. Calibration of ^{14}C dates of individuals from the necropolis associated to the Santa María la Real, obtained with OxCal v 4.1 (Bronk Ramsey, 2009) and IntCal09 atmospheric data (Reimer et al., 2009).

5. Conclusions

Lime mortar radiocarbon dating will result ambiguous because of CO_2 contamination from inherited carbonates and, burial and/or weathering modifications. Therefore, an effective extraction method of binder is necessary in order to prevent the determination of *older* or *younger* ages.

A new combined mechanical and physical procedure was established to obtain very pure binder fraction. The extraction of particles of less than 1 μm in size was developed allowing to achieve suitable material for lime mortar dating. This extraction method eliminates the sources of contamination in carbon, fundamental issue for radiometric dating of lime mortars. Thus, this procedure removes partial acid digestion from dating procedure of binder and several radiocarbon measurements of complex interpretation. The obtained pure binder after separation procedure is enough in quantity and quality for radiocarbon dating. The adequate binder carbonate selection has a broad application on establishing the chronology of the different construction phases of buildings.

For the selection of a suitable sample for radiocarbon dating, petrographic analysis is essential. The identification of different mineralogical phases of mortar samples, aggregate nature, and the limestone fragment remains as well as the recognition of possible features of mortar degradation (organic and inorganic) are issues to elucidate. Moreover, analysis performed by Scanning Electron Microscopy allows to identify contaminant error sources of the mortar sample on a small scale.

The effectiveness of mechanical separation and the verification of binder purity have been proved by X-ray diffraction, Scanning Electron Microscope analyses and thermogravimetric analysis and differential thermal analysis.

The preparation method was successfully applied to distinctive physical characteristic bearing and varying age five mortars. Mortars correspond to perimetric walls of different age buildings of the Santa María la Real church in Zarautz. Moreover, the elimination of the source of error in inherited carbon provides ages that are relatively insensitive to aging when determining the dates of these historic buildings and of their construction phases.

The radiometric ages of studied mortars are consistent with the established by archaeological and historical studies for different construction phases of the studied church. Therefore, the procedure of sample preparation considerably simplifies the performance of radiometric measurements and their interpretation.

6. Acknowledgments

This paper has received financial support from the project IT-315-10 of the Basque Government and UNESCO 09/01 and EHU10/32 projects of The University of the Basque Country. We thank Christine Laurin, who has greatly improved the English text.

7. References

Adriano, P., Silva, A. S., Veiga, R., Mirao, J. & Candeias, A. E. (2009). *Microscopic characterisation of old mortars from the Santa Maria Church in Evora*, Materials Characterization, Vol. 60, No. 7, pp. 610-620, ISSN 1044-5803

Al-Bashaireh, K. & Hodgins, G. W. L. (2011). *AMS 14C dating of organic inclusions of plaster and mortar from different structures at Petra-Jordan*, Journal of Archaeological Science, Vol. 38, No. 3, pp. 485-491, ISSN 0305-4403

Ambers, J. (1987). *Stable carbon isotope ratios and their relevance to the determination of accurate radiocarbon-dates for lime mortars*, Journal of Archaeological Science, Vol. 14, No. 6, pp. 569-576, ISSN 0305-4403

Bakolas, A., Biscontin, G., Moropoulou, A. & Zendri, E. (1998). *Characterization of structural byzantine mortars by thermogravimetric analysis*, Thermochimica Acta, Vol. 321, No. 1-2, pp. 151-160, ISSN 0040-6031

Ball, R. J., El-Turki, A. & Allen, G. C. (2011). *Influence of carbonation on the load dependent deformation of hydraulic lime mortars*, Materials Science and Engineering: A, Vol. 528, No. 7-8, pp. 3193-3199, ISSN 0921-5093

Baxter, M. S. & Walton, A. (1970). *Radiocarbon dating of mortars*, Nature, Vol. 225, No. 5236, pp. 937-938, ISSN 0028-0836

Benea, V., Vandenberghe, D., Timar, A., Van den Haute, P., Cosma, C., Gligor, M. & Florescu, C. (2007). *Luminescence dating of Neolithic ceramics from Lumea Noua, Romania,* Geochronometria, Vol. 28, pp. 9–16, ISSN 1733-8387

Berger, R. (1992). *C-14 dating mortar in Ireland,* Radiocarbon, Vol. 34, No. 3, pp. 880-889, ISSN 0033-8222

Bowman, S., (1990). *Radiocarbon dating,* University of California Press ISBN 0520070372, Berkeley

Bronk Ramsey, C. (2009). *Bayesian analysis of radiocarbon dates,* Radiocarbon, Vol. 51, No. 1, pp. 337-360, ISSN 0033-8222

Casadio, F., Chiari, G. & Simon, S. (2005). *Evaluation of binder/aggregate ratios in archaeological lime mortars with carbonate aggregate: A comparative assessment of chemical, mechanical and microscopic approaches,* Archaeometry, Vol. 47, pp. 671-689, ISSN 0003-813X

Ceberio, M., (2009). La cerámica común no torneada de época romana del yacimiento de Santa María la Real de Zarautz (País Vasco). Una aproximación a su caracterización tipológica, *in Santa María La Real de Zarautz (País Vasco) continuidad y discontinuidad en la ocupación de la costa vasca entre los siglos V a. C. y XIV d. C,* Ibáñez Etxeberria, Á. (Ed.), pp. 176-190, Sociedad de Ciencias Aranzadi ISBN 978-84-931930-9-6, Donostia

Cepeda Ocampo, J. J., (2009). Hallazgos romanos en Santa María la Real de Zarautz (País Vasco). La terra sigilata, las lucernas y monedas, *in Santa María La Real de Zarautz (País Vasco) continuidad y discontinuidad en la ocupación de la costa vasca entre los siglos V a. C. y XIV d. C,* Ibáñez Etxeberria, Á. (Ed.), pp. 258-272, Sociedad de Ciencias Aranzadi ISBN 978-84-931930-9-6, Donostia

Cimitan, L., P., R. & Zaninetti, A. (1991). *Studio delle tecniche di disgregazione per le indagini diagnostiche delle malte. Materiali e Strutture,* Problemi di Conservazione, Vol. 3, pp. 121–130, ISSN 1121-2373

Corazza, M., Pratesi, G., Cipriani, C., Lo Guidice, A., Rossi, P., Vittone, E., Manfredotti, C., Pecchioni, E., Manganelli del Fa, C. & Fratini, F. (2001). *Ionoluminescence and cathodoluminescence in marbles of historic and architectural interest,* Archaeometry, Vol. 43, pp. 439-446, ISSN 0003-813X

Davis, J. A. & Kent, D. B., (1990). Surface complexation modeling in aqueous geochemistry, *in Mineral-Water Interface Geochemistry,* Hochella, M.F. & White, A.F. (Eds.), pp. 177-260, Mineralogical Society of America ISBN 0-939950-28-6,

Esteban Delgado, M. (2004). *Tendencia en la creacción de asentamientos durantes los primeros siglos de la era en el espacio litoral guipuzcuano,* Kobie, Vol. 6, No. 1, pp. 371-380, ISSN

Esteban Delgado, M., Martínez Salcedo, A., Ortega Cuesta, L. A., Alonso-Olazabal, A., Izquierdo Marculeta, M. T., Rechin, F. & Zuluaga Ibargallartu, M. C., (in press). *Caracterización tecnológica y arqueológica de la cerámica común no torneada de época romana en el País Vasco peninsular y Aquitania meridional: Producción, difusión, funcionalidad, cronología,* Eusko Ikaskuntza, Donostia

Fernández Ochoa, C. & Morillo Cerdán, Á., (1994). *De Brigantium a Oiasso : una aproximación al estudio de los enclaves maritimos cantábricos en época romna,* Foro Arqueología Proyectos y Publicaciones, ISBN 8460501051, Madrid

Folk, R. L. & Valastro, S. J. (1976). *Successful technique for dating of lime mortar by carbon-14,* Journal of Field Archaeology, Vol. 3, pp. 203–208., ISSN 0093-4690

Frumkin, A., Shimron, A. & Rosenbaum, J. (2003). *Radiometric dating of the Siloam Tunnel, Jerusalem,* Nature, Vol. 425, No. 6954, pp. 169-171, ISSN 0028-0836

Goedicke, C. (2003). *Dating historical calcite mortar by blue OSL: results from known age samples,* Radiation Measurements, Vol. 37, No. 4-5, pp. 409-415, ISSN 1350-4487

Goslar, T., Nawrocka, D. & Czernik, J. (2009). *Foraminiferous limestone in C-14 dating of mortar*, Radiocarbon, Vol. 51, No. 3, pp. 987-993, ISSN 0033-8222

Habermann, D., Neuser, R. & Richter, K., (2000). Quantitative high resolution spectral analysis of Mn2+ in sedimentary calcite, *in Cathodoluminescence in geosciences*, Pagel, M. (Ed.), pp. 331-358, Springer ISBN Berlin

Hale, J., Heinemeier, J., Lancaster, L., Lindroos, A. & Ringbom, Å. (2003). *Dating ancient mortar*, American Scientist, Vol. 91, pp. 130–137, ISSN 0003-0996

Heinemeier, J., Jungner, H., Lindroos, A., Ringbom, s., von Konow, T. & Rud, N. (1997a). *AMS 14C dating of lime mortar*, Nuclear Instruments and Methods in Physics Research Section B: Beam Interactions with Materials and Atoms, Vol. 123, No. 1-4, pp. 487, ISSN 0168-583X

Heinemeier, J., Jungner, H., Lindroos, A., Ringbom, s., von Konow, T., Rud, N. & Sveinbjornsdottir, A., (1997b). AMS C-14 dating of lime mortar, *in Proceedings of the VII Nordic Conference on the Application of Scientific Methods in Archaeology*, Edgren, T. (Ed.), pp. 214-215, ISBN 951-9057-27-7,

Heinemeier, J., Ringbom, A., Lindroos, A. & Sveinbjornsdottir, A. E. (2010). *Successful AMS C-14 dating of non-hydraulic lime mortars from the medieval churches of the Aland Islands, Finland*, Radiocarbon, Vol. 52, No. 1, pp. 171-204, ISSN 0033-8222

Hiekkanen, M., (1998). Finland's medieval stone churches and their dating - a topical problem., Helsinki, Suomen Museo, p. 143-149.

Ibañez Etxeberria, A., (2009). *Santa María la Real de Zarautz (País Vasco): continuidad y discontinuidad en la ocupación de la costa vasca entre los siglos V a.C. y XIV d.C*, Sociedad de Ciencias Aranzadi ISBN 978-84-931930-9-6, Donostia

Ibañez Etxeberria, A. & Sarasola Etxegoien, N., (2009). El yacimiento arqueológico de Santa María de Zarautz (País Vasco), *in Santa María la Real de Zarautz (País Vasco): continuidad y discontinuidad en la ocupación de la costa vasca entre los siglos V a.C. y XIV d.C*, Ibáñez Etxeberria, Á. (Ed.), pp. 12-84, Sociedad de Ciencias Aranzadi ISBN 978-84-931930-9-6, Donostia

Ibáñez Etxeberria, A., (2003). *Entre Menosca e Ipuscua: arqueología y territorio en el yacimiento de Santa María La Real de Zarautz (Gipuzkoa)*, Zarauzko Arte eta Historia Museoa, ISBN 8492303344, Zarautz

Ibáñez Etxeberria, A. & Moraza, A. (2005). *Evolución cronotipológica de las inhumaciones medievales en el Cantábrico Oriental: el caso de Santa María la Real de Zarautz (Gipuzkoa)*, Munibe, Suplemento, Vol. 57, pp. 419-437, ISSN 1132-2217

Kosednar-Legenstein, B., Dietzel, M., Leis, A. & Stingl, K. (2008). *Stable carbon and oxygen isotope investigation in historical lime mortar and plaster Results from field and experimental study*, Applied Geochemistry, Vol. 23, No. 8, pp. 2425, ISSN

Labeyrie, J. & Delibrias, G. (1964). *Dating of old mortar by Carbon-14 method*, Nature, Vol. 201, No. 492, pp. 742-743, ISSN 0028-0836

Laird, D. A. & Dowdy, R. H., (1994). Preconcentration techniques in soil mineralogical analyses, *in Quantitative Methods in Soil Mineralogy*, Luxmoore, R.J. (Ed.), pp. 236-266, Soil Science Society of America ISBN 0-89118-806-1, Madison,WI

Lanas, J., Sirera, R. & Alvarez, J. I. (2005). *Compositional changes in lime-based mortars exposed to different environments*, Thermochimica Acta, Vol. 429, No. 2, pp. 219-226, ISSN 0040-6031

Lindroos, A., Heinemeier, J., Ringbom, Å., Braskén, M. & Sveinbjörnsdóttir, Á. (2007). *Mortar dating using AMS 14C and sequential dissolution: examples from medieval, non-hydraulic lime mortars from the Åland Islands, SW Finland*, Radiocarbon, Vol. 49, No. 1, pp. 47–67, ISSN 0033-8222

Machel, H. G., (2000). *Application of cathodoluminescence to carbonate diagenesis*, ISBN 3-540-65987-0,

Marques, S. F., Ribeiro, R. A., Silva, L. M., Ferreira, V. M. & Labrincha, J. A. (2006). *Study of rehabilitation mortars: Construction of a knowledge correlation matrix*, Cement and Concrete Research, Vol. 36, No. 10, pp. 1894-1902, ISSN 0008-8846

Marshall, D. J., (1988). *Cathodoluminescence of geological materials*, Allen & Unwin, ISBN 0045520267 (alk. paper), Boston

Marzaioli, F., Lubritto, C., Nonni, S., Passariello, I., Capano, M. & Terrasi, F. (2011). *Mortar Radiocarbon Dating: Preliminary Accuracy Evaluation of a Novel Methodology*, Analytical Chemistry, Vol. 83, No. 6, pp. 2038-2045, ISSN 0003-2700

Nawrocka, D., Michniewicz, J., Pawlyta, J. & Pazdur, A. (2005). *Application of radiocarbon method for dating of lime mortars*, Geochronometria, Vol. 24, pp. 109-115, ISSN 1733-8387

Nawrocka, D., Czernik, J. & Goslar, T. (2009). *C-14 dating of carbonate mortars from Polish and Israeli sites*, Radiocarbon, Vol. 51, No. 2, pp. 857-866, ISSN 0033-8222

Nawrocka, D. M., Michczynska, D. J., Pazdur, A. & Czernik, J. (2007). *Radiocarbon chronology of the ancient settlement in the Golan Heights area, Israel*, Radiocarbon, Vol. 49, No. 2, pp. 625-637, ISSN 0033-8222

Ortega, L. A., Zuluaga, M. C., Alonso-Olazabal, A., Insausti, M. & Ibañez, A. (2008). *Geochemical characterization of archaeological lime mortars: Provenance inputs*, Archaeometry, Vol. 50, pp. 387-408, ISSN 0003-813X

Paama, L., Pitkanen, I., Ronkkomaki, H. & Peramaki, P. (1998). *Thermal and infrared spectroscopic characterization of historical mortars*, Thermochimica Acta, Vol. 320, No. 1-2, pp. 127-133, ISSN 0040-6031

Pachiaudi, C., Marechal, J., Van Strydonck, M., Dupas, M. & Dauchotdehon, M. (1986). *Isotopic fractionation of carbon during CO2 absorption by mortar*, Radiocarbon, Vol. 28, No. 2A, pp. 691-697, ISSN 0033-8222

Pagel, M., (2000). *Cathodoluminescence in geosciences*, Springer, ISBN 3540659870, Berlin

Rech, J. A., Fischer, A. A., Edwards, D. R. & Jull, A. J. T. (2003). *Direct dating of plaster and mortar using AMS radiocarbon: A pilot project from Khirbet Qana, Israel*, Antiquity, Vol. 77, No. 295, pp. 155-164, ISSN 0003-598X

Rech, J. A. (2004). *New uses for old laboratory techniques*, Near Eastern Archaeology, Vol. 67, No. 4, pp. 212-219, ISSN 1094-2076

Reimer, P. J., Baillie, M. G. L., Bard, E., Bayliss, A., Beck, J. W., Blackwell, P. G., Ramsey, C. B., Buck, C. E., Burr, G. S., Edwards, R. L., Friedrich, M., Grootes, P. M., Guilderson, T. P., Hajdas, I., Heaton, T. J., Hogg, A. G., Hughen, K. A., Kaiser, K. F., Kromer, B., McCormac, F. G., Manning, S. W., Reimer, R. W., Richards, D. A., Southon, J. R., Talamo, S., Turney, C. S. M., van der Plicht, J. & Weyhenmeye, C. E. (2009). *IntCal09 and Marine09 radiocarbon age calibration curves, 0-50,000 years cal BP*, Radiocarbon, Vol. 51, No. 4, pp. 1111-1150, ISSN 0033-8222

Salama, A. I. A. & Ian, D. W., (2000). Mechanical techniques: particle size separation, *in Encyclopedia of Separation Science* pp. 3277-3289, Academic Press ISBN 978-0-12-226770-3, Oxford

Schmid, S. G. (2001). *The International Wadi Farasa Project IWF, preliminary report on the 1999 season*, The Annual of the Department of Antiquities of Jordan, Vol. 45, pp. 343–357, ISSN

Sickels, L. B., (1981). Organics vs. synthetics: their use as additives in mortars: *Symposium on Mortars, Cements and Grouts Used in the Conservation of Historic Buildings*, pp. 25–53.

Sonninen, E., Erametsa, P. & Jungner, H., (1989). Dating of mortar and bricks: an example from Finland: *Archaeometry: proceedings of the 25th international symposium* pp. 99-107.

Sonninen, E. & Jungner, H. (2001). *An improvement in preparation of mortar for radiocarbon dating*, Radiocarbon, Vol. 43, No. 2A, pp. 271-273, ISSN 0033-8222

Soukup, D. A., Buck, B. J. & Harris, W., (2008). Preparing soils for mineralogical analyses, *in Methods of Soil Analysis. Part 5 – Mineralogical Methods*, Ulery, A.L. & Drees, L.R. (Eds.), pp. 12-31, Soil Science Society of America, Inc ISBN 978-0-89118-846-9, Madison

Stefanidou, M. & Papayianni, I. (2005). *The role of aggregates on the structure and properties of lime mortars*, Cement and Concrete Composites, Vol. 27, No. 9-10, pp. 914-919, ISSN 0958-9465

Stokes, G. G. (1851). *On the effect of the lateral friction of fluids on the motion of pendulums*, Trans. Cambridge Phil. Soc., Vol. 9, pp. 8-108, ISSN

Stuiver, M. & Smith, C. S., (1965). Radiocarbon dating of ancient mortar and plaster: *Proceedings of the 6th International 14C Conference*, pp. 338-343.

Tubbs, L. E. & Kinder, T. N. (1990). *The use of AMS for the dating of lime mortars*, Nuclear Instruments and Methods in Physics Research Section B: Beam Interactions with Materials and Atoms, Vol. 52, No. 3-4, pp. 438, ISSN

Van Strydonck, M., Dupas, M. & Dauchot-Dehon, M., (1983). Radiocarbon dating of old mortars: *14C and Archaeology, Proceedings*, pp. 337–343.

Van Strydonck, M., Dupas, M., Dauchotdehon, M., Pachiaudi, C. & Marechal, J. (1986). *The influence of contaminating (fossil) carbonate and the variations of delta-C-13 in mortar dating*, Radiocarbon, Vol. 28, No. 2A, pp. 702-710, ISSN 0033-8222

Van Strydonck, M. & Dupas, M., (1991). The classification and dating of lime mortars by chemical analysis and radiocarbon dating: a review, *in Second Deya International Conference of Prehistory : recent developments in western Mediterranean prehistory: archaeological techniques, technology, and theory*, Waldren, W.H., Ensenyat, J.A. & Kennard, R.C. (Eds.), pp. 5-43, Tempus Reparatum ISBN 0860547272, Oxford

Van Strydonck, M. J. Y., Van der Borg, K., De Jong, A. F. M. & Keppens, E. (1992). *Radiocarbon dating of lime fractions and organic material from buildings*, Radiocarbon, Vol. 34, pp. 873–879, ISSN 0033-8222

Warkentin, B. P. & Maeda, T., (1980). Physical and mechanical characteristics of andisols *in Soils with Variable Charge*, Theng, B.K.G. (Ed.), pp. 281-302, New Zealand Society of Soil Science ISBN Lower Hutt.

Wilson, R. & Spengler, J. D., (1996). *Particles in our air : concentrations and health effects*, Harvard University Press, ISBN 0674240774, Cambridge, Mass

Wintle, A. G. (2008). *Fifty years of luminescence dating**, Archaeometry, Vol. 50, No. 2, pp. 276-312, ISSN 1475-4754

Wyrwa, A. M., Goslar, T. & Czernik, J. (2009). *AMS C-14 dating of romanesque rotunda and stone buildings of a medieval monastery in Lekno, Poland*, Radiocarbon, Vol. 51, No. 2, pp. 471-480, ISSN 0033-8222

Zamba, I. C., Stamatakis, M. G., Cooper, F. A., Themelis, P. G. & Zambas, C. G. (2007). *Characterization of mortars used for the construction of Saithidai Heroon Podium (1st century AD) in ancient Messene, Peloponnesus, Greece*, Materials Characterization, Vol. 58, No. 11-12, pp. 1229-1239, ISSN 1044-5803

2

Dating of Old Lime Based Mixtures with the "Pure Lime Lumps" Technique

Giovanni L.A. Pesce and Richard J. Ball
BRE Centre for Innovative Construction Materials,
Department of Architecture and Civil Engineering,
University of Bath, Bath,
United Kingdom

1. Introduction

A number of studies carried out over the last forty years describe the application of radiocarbon dating of lime mixtures such as mortars, plasters and renders (Folk & Velastro, 1976, Pachiaudi et al., 1986, Van Strydonck et al., 1992, Berger, 1992, Hale et al., 2003, Sonninen & Junger, 2001, Nawrocka et al., 2005).

To understand the basic principle of this technique we must first consider the lime manufacturing process. Lime is produced from limestone (essentially calcium carbonate of geological origin) which is burnt at around 900°C to produce carbon dioxide and calcium oxide otherwise known as "quicklime" (Norman, 1961, Oates, 1998, Vecchiattini, 2010, Rattazzi, 2007, Goren & Goring-Morris 2008).

The quicklime subsequently undergoes a slaking process where it is mixed with water. The resulting exothermic reaction forms calcium hydroxide which, in the past, was mainly produced using an excess of water which resulted in the formation of "lime putty", one of the most widely used inorganic binders for construction. Plasters, renders, mortars and other architectural parts are formed when the lime putty is mixed with a sand (aggregate).

When in place the calcium hydroxide within the mixtures hardens by carbonation resulting in the formation of new calcium carbonate. As the carbon that binds to the calcium during this reaction originates from the atmosphere, the calcium carbonate contained in the hardened mixture reflects the atmospheric radiocarbon concentration at the time of hardening. Consequently this calcium carbonate phase within lime mixtures can be used for the radiocarbon dating of mortars, plasters and other lime based materials.

Despite the fact that this method is very simple in principle, several studies have highlighted various practical challenges and factors that must be considered (Van Strydonck et al., 1986). These arise mainly from the contamination of samples with carbonaceous substances such as incompletely burnt limestone and aggregates of fossil origin including limestone sand.

However, recently studies have shown that accurate sample processing allow a significant reduction of these error sources (Sonninen & Junger, 2001, Marzaioli et al., 2011 and see also the contribution of Ortega and its colleagues in this book) and moreover adoption of a special sampling procedure based on the careful selection of lumps of incompletely mixed lime, provides an interesting alternative that avoids problems associated with contamination.

This latter technique has been discussed at an international level on several occasions (e.g. Gallo et al., 1998) and the more recent publications on this subject describe the results achieved using the latest sampling methodologies, together with the limitations and drawbacks (Pesce, 2010).

The founding principle underlying this technique is the use of the pure lime lumps. These are thought to originate from imperfect mixing and are most prevalent in mortars, renders and plasters predating mechanical mixing. Previous sampling methods for radiocarbon dating did not discriminate between pure and contaminated lime lumps. As pure lumps contain the same lime as that used in other parts of the mixtures but importantly are free of contaminants such as sand grains or under burned pieces of limestone, they can dramatically reduce the errors in the radiocarbon dating.

In the following paragraphs more details of this technique are given together with examples of some applications.

2. The radiocarbon dating of lime based materials: Basic principles

2.1 Carbonation and radiocarbon in building materials

Carbonation is one of the main reactions (if not the main reaction) of lime based materials that lead to hardening and, consequently, to an increase in strength (Van Balen, 2005, El-Turki et al., 2007).

From a chemical point of view, carbonation is the reaction between calcium ions (Ca^{2+}) and carbonate ions (CO_3^{2-}) in aqueous solution. This leads to the precipitation of a solid calcium carbonate phase, as described by the following reaction scheme:

- dissolution of carbon dioxide (CO_2) in water (Morse & Mackenzie, 1990):

$$CO_{2(g)} \leftrightarrow CO_{2(aq)} \tag{1}$$

$$CO_{2(aq)} + H_2O_{(l)} \leftrightarrow H_2CO_{3(aq)} \tag{2}$$

$$H_2CO_{3(aq)} \leftrightarrow HCO_3^-{}_{(aq)} + H^+{}_{(aq)} \tag{3}$$

$$HCO_3^-{}_{(aq)} \leftrightarrow CO_3^{2-}{}_{(aq)} + H^+{}_{(aq)} \tag{4}$$

- dissolution of calcium hydroxide ($Ca(OH)_2$) in water:

$$Ca(OH)_{2(s)} + H_2O_{(l)} \leftrightarrow Ca^{2+}{}_{(aq)} + 2(OH^-)_{(aq)} + H_2O_{(l)} \tag{5}$$

- reaction of calcium ions with carbonate ions in water solution:

$$Ca^{2+}{}_{(aq)} + CO_3^{2-}{}_{(aq)} + H_2O_{(l)} \leftrightarrow CaCO_{3(s)} + H_2O_{(l)} \tag{6}$$

These reactions describe the dissolution of atmospheric carbon dioxide (CO_2) in water (reaction 1) and the subsequent reaction forming carbonic acid (2). Once in solution, carbonic acid dissociates forming bicarbonate (HCO_3^-) and a hydrogen ion (H^+; reaction 3). Subsequently, this bicarbonate ion may dissociate to form a carbonate ion and a further hydrogen ion (4). Concurrently, calcium hydroxide ($Ca(OH)_2$) dissolves in water forming calcium and hydroxyl ions (OH^-; 5).

Depending on several factors such as temperature, ion concentration and the presence of other ions in solution, a solid calcium carbonate phase precipitate (6).

Inspection of reactions 1-4 and 6 illustrates that calcium carbonate crystals precipitated from a saturated solution of lime in water (6) contain carbon atoms present in the air at the time of the reactions (1). Thus the radiocarbon analysis records the carbon contained in the air at the time of the crystal precipitation.

Moreover, as the carbon atoms in atmospheric carbon dioxide exist in at least three main isotopic forms (^{12}C, ^{13}C, ^{14}C), the calcium carbonate precipitated during hardening will contain these three forms in a similar ratio to that of the atmosphere at the time of precipitation[1].

An important consequence is that radiocarbon is incorporated within the $CaCO_3$ crystals and begins to decay from the time of precipitation. As no carbon exchange can happen in the solid phase of calcium carbonate, ^{14}C contained in lime based mortars, plasters and renders can be used to date these materials with the radiocarbon method.

Furthermore, if we assume that the time of calcium carbonate precipitation is the same as that of the buildings construction, where the lime has been used, it follows that the radiocarbon technique can be used to date old structures such as buildings, bridges and churches.

2.2 Drawbacks and limitations of radiocarbon dating of lime mixtures

Despite the simplicity of the carbonation process previously described, in practice the radiocarbon dating of lime mixtures is not so simple. Several studies have recognized the drawbacks and limitations of radiocarbon dating of lime based materials. These are mainly attributed to contamination of samples from carbonaceous substances such as incompletely burnt limestone or grains of carbonate sand.

Carbonate rocks are in fact mainly made of calcium carbonate that, at the beginning of the sedimentary process, contained all the carbon isotopes contained in the earths' atmosphere at the time of the precipitation. However, in contrast to the carbon contained in the lime mixtures of archaeological interest which have a lifetime shorter than the radiocarbon, calcium carbonate of geological origin comes from times far earlier than that equivalent to the decay time of radiocarbon. This means that any isotope contained in carbonate rocks is "^{14}C-dead".

---

[1]This is only an approximation because of the kinetic fractionation of isotopes during the calcium carbonate precipitation. An example of this fractionation is described, among the other scientific references, in: Turner, 1982 or Emrich et al., 1970.

Importantly, this means that any piece of carbonate rock that may contaminate a hardened lime such as grains of carbonate sand added as aggregate to the mixture or pieces of under burned limestone[2] can affect the result of radiocarbon dating.

Presence of aggregate or pieces of stones would not be a problem if it were possible to separate these phases from the binder phase in lime mixtures. But, even if under the microscope it is possible to distinguish between binder (lime) and aggregate (sand), from a practical point of view, it is almost impossible to separate all the aggregate from the binder when both are made of calcium carbonate.

In fact, if sand contained in the mixtures is only made of silica, there are already available chemical procedures to separate the calcium carbonate of the binder from that of the aggregate. However, this case is not frequent and, moreover, there still remains the problem of under burned pieces of stone that behave in a similar manner of the binder under acid attack.

Improved methodologies have been developed to allow similar procedures with mixtures containing carbonate sand but, despite this, there still remains some uncertainty in the radiocarbon dating of these types of mortar (Folk & Velastro, 1976, Van Strydock et al., 1986, Sonninen & Junger, 2001).

Looking more carefully to the limitations of radiocarbon dating of lime based mixtures, it is noteworthy that the previous statement about the immutability of carbon contained in calcium carbonate crystals is subject to an exception. Sometimes, in fact, lime contained in building materials can be affected by a dissolution and re-precipitation process of calcium carbonate that can lead to an exchange of carbon atoms.

Typical water sources of archaeological and building sites such as rain (mainly for renders), backwater or rising dump (for plasters, renders, and mortars), can lead to the dissolution of calcium carbonate. Calcium carbonate has a specific solubility in water[3] and, consequently, binder contained in the mixtures can dissolve in very wet environments such as foundations and underground floors. Binder dissolution in old lime based mixtures can be described as follows:

$$CaCO_{3(s)} + H_2O_{(l)} \leftrightarrow Ca2^{+}_{(aq)} + CO_3^{2-}{}_{(aq)} + H_2O_{(l)} \qquad (7)$$

Once in solution calcium ions can react with carbonate ions in the manner already described (equation 6), to produce new calcium carbonate crystals. However, in this case the carbonate ion population dissolved in solution can be made of both, ions from the carbonate dissolution and ions from the atmosphere. But carbon concentration in the atmosphere at the dissolution time of the lime can be different from the carbon concentration at the hardening time of the original lime.

For this reason, the radiocarbon dating of a lime sample containing even a small amount of re-precipitated calcium carbonate can be affected by errors due to the radiocarbon contents of the new calcium carbonate phases.

---

[2]Pieces of the same stone used to produce the lime that have not been completely decomposed during the productive process.

[3]Even if the solubility of calcium carbonate is less than other building materials such as gypsum.

To avoid this problem it is necessary to analyse the sample used for the dating before submission to the laboratory were the radiocarbon counting is carried out. Re-precipitated calcium carbonate in lime mixtures can be recognised under the microscope with the cathodoluminescence technique (Gliozzo & Memmi Turbanti, 2006) or by other more common techniques such as X ray diffraction (XRD) and infrared spectroscopy (FT-IR).

Finally, to assume that the hardening time of lime is the same as the structure construction time is only an approximation, sometimes not verified.

Carbonation is a very slow process that depends on several factors such as temperature, moisture content and pore structure which dictates the accessibility of CO_2. As all these factors are quite changeable from site to site and also inside the same site, the hardening time of lime mixtures can be variable.

Carbonation can stop if the environment where the reaction occurs is too dry or if the CO_2 access is inhibited (Van Balen, 2005). Under these conditions carbonation can only start again following a change in the conditions such as a break in the wall or a small fracture in the mortar due to shrinkage.

Investigations of old buildings have already highlighted that the inner part of thick walls, such as the city walls, can contain calcium hydroxide centuries after their construction. This means that the results of radiocarbon dating of old lime mixtures depend not only on the purity of the sample itself, but also on the depth of the sample within the analysed structure.

For this reason, if the radiocarbon dating of a sample of lime mortar is carried out to date the building process of the wall where the sample was picked up, a superficial sample is more desirable than a deep sample.

But in this case the sampling procedure should avoid lime applied to the walls after their construction such as joint sealing whose date of origin postdates the building time of the wall.

Renders and plasters should not be affected by the problem of delayed carbonation because they are in direct contact with the atmosphere and their structure is porous thereby permitting diffusion of carbon dioxide. However, in this case, it must be verified that these layers, often considered expendable materials, have not been subject to demolition and reconstruction. If this has happened, obviously, their radiocarbon dating can not be assumed as the radiocarbon dating of the wall but simply as the date of the re-plastering.

3. The pure lime lumps technique

3.1 Advantages in the use of lime lumps

Summarising the problems previously listed, it is possible to state that errors in the radiocarbon dating of old lime mixtures mainly arise from three factors:

- re-carbonation of the lime binder;
- delay in carbonation;
- contamination from external sources of calcium carbonate.

Errors resulting from the first two types of problem can be easily mitigated by correct sample selection and accurate analysis before the radiocarbon dating. The third type of error, instead, can be avoided by applying the "pure lime lumps technique". That is selecting a specific part of the mixture where lumps of pure lime are found embedded.

It is known that during the older manufacturing process of lime mixtures, some lumps of lime putty not mixed with sand, could remain embedded within the mixture matrix.

A feature of this process was that when the pieces of burned limestone (calcium oxide) were removed from the furnaces, they were immersed in pools containing water where the slaking process took place forming lime putty (i.e. calcium hydroxide). Slaking was assisted by a continuous manual mixing carried out with traditional tools. Once the quick lime was completely dissolved, the lime putty was then filtered and stored in pits dug in the ground or immediately mixed with sand and other aggregates to produce mixtures such as mortars, plasters and renders (Oates 1998, Vecchiattini, 2010, Rattazzi 2007).

However, during this process, some small lumps of lime putty could remain compact and fail to intimately mix with the sand grains (probably because they were not well slaked or mixed[4]) even though they were able to combine with the atmospheric CO_2 in the same manner as the remaining matrix. According to results of research into this topic the composition of these lime lumps is similar to that of the surrounding matrix (Franzini et al., 1990, Bruni et al., 1997, Bakolas et al., 1995) with only differences in micro-morphology[5]. These differences are described by Bakolas and colleagues as: "*The matrix of the mortar surrounding the lumps appear to be of compact texture with smaller porosity in respect to the lumps*" and, moreover, "*in many cases the growth of the binder crystals is greater than of the lumps*" (Bakolas et al., 1995, p. 814). The same observations have been made by other researchers including as Bugini and Toniolo (Bugini & Toniolo, 1990).

For this reason the lumps of pure lime embedded in old lime mixtures are the most suitable samples for the radiocarbon dating of lime based materials. With their use the contamination problem previously described is greatly reduced. The only contaminants remain on the external surface where the lump was originally jointed with the mortar matrix and these are easily removable.

3.2 Lumps in old lime based mixtures

For successful application of the "pure lime lumps" technique it is important to identify these lumps and distinguish them from several other types commonly found embedded within old mixtures. The internal structure of historic lime mixtures has been found to contain at least five different types of lump, each of which can be recognised individually. These comprise:

- unburned pieces of limestone (Leslie & Hughes, 2002, Ingham, 2005, Elsen, 2006);
- over burned pieces of limestone (Leslie & Hughes 2002, Ingham, 2005, Elsen, 2006, Elsen et al., 2004);

[4]There is not a general agreement about the formation process of these lumps but the above mentioned hypothesis is one of the most common (Bugini & Toniolo, 1990).

[5]For this reason, some authors hypothesize that crystals in these lumps developed over a shorter time period compared to the crystals of the matrix (Bruni et al., 1997).

- pieces of burned limestone containing high concentrations of silica (these arise when the stone used for the lime production contains high quantity of impurities; Elsen et al., 2004, Bakolas et al., 1995);
- concretions of recarbonated lime;
- lumps of pure calcium carbonate due to the carbonation of lime putty (Leslie & Hughes, 2002, Franzini et al., 1990, Bugini & Toniolo, 1990, Ingham, 2005, Elsen, 2006, Elsen, et al., 2004, Bakolas et al., 1995).

Among these only the lumps belonging to the latter group are suitable for radiocarbon dating. Over-burned pieces of lime contain sintered calcium oxide which is less reactive with water (Elsen, 2006). Consequently, if carbon is contained within these samples, it must not be considered representative of the atmospheric carbon dioxide at the time of the building process. Unburned pieces of limestone contain carbon dioxide of geological origin, while calcium contained inside the lumps of silica is mainly bonded to the silicon rather than to the carbon dioxide (Bakolas et al., 1995). Problem connected to the concretions have already been discussed previously (par. 2.2)

3.3 Recognition and sampling of the pure lime lumps

Lumps of pure lime are easily identifiable in old lime mixtures as they exhibit a white, rounded and floury complexion (Fig. 1). Surface hardness of this type of lump is very low making them extremely delicate to handle and easily damaged. To facilitate extraction of these lumps in an undamaged condition a two-step procedure is used:

1. on site sampling of a small amount of mortar containing the lump;
2. extraction of suitable pure lime lumps for the laboratory analysis.

Fig. 1. Pure lime lumps in a specimen of air lime mortar (left hand side; scale bar: 1 cm) and detail of the lime lump (right hand side)

3.3.1 The on site sampling procedure

On-site sampling must be tailored to reduce problems arising from the re-carbonation of lime described previously. Moreover, sampling should be carried out to accommodate the individual requirements of the structure from which the sample is taken. Buildings out of the ground, underground walls, frescos layers, mosaic substrate, all require slight variations in technique.

In the field of building archaeology, for instance, it can be difficult to reach the inner part of walls, especially if the thickness of mortar joints is not large enough to allow selection of suitable samples[6].

When the inner part of the masonry is accessible, it is always important to consider the possible re-carbonation of lime lumps and the depth from which samples within the wall should be taken. Care is also needed to avoid unusual situations such as water pockets and mixtures non belonging to the original structure such as pieces of plaster applied after its construction.

In the same manner samples taken deep inside the wall, where incompletely or delayed carbonated lime may be present, should be avoided. In this case it is useful to remember that the carbonation process initiates from the external surfaces of a structure and progresses towards the inner region at decreasing speed.

When a suitable depth of sampling is reached, a lump containing sufficient material for the radiocarbon dating must be identified. Where an Accelerator Mass Spectrometer (AMS) is used in the radiocarbon dating, at least 20 milligrams of calcium carbonate will be required.

In the case that a single lump does not contain sufficient material, multiple lumps from the same region of the masonry can be used (in archaeological terms this means from the same stratigraphic unit).

3.3.2 Sample extraction inside the laboratory

Following on-site sampling and before treatment and analysis at the AMS laboratory, it is necessary to examine the samples under an optical stereo microscope to confirm the lump nature and mechanically remove particles of aggregate that may be still attached to the surface.

On-site it is difficult to distinguish lumps of incompletely burnt limestone or rounded grains of milky quartz from pure lime lumps. However under a magnifying glass, even at low magnification, it is possible to distinguish between these types. The surfaces of lime lumps have a floury appearance (Fig. 1 right hand side) while those of under burned lumps and sand grains appear denser resembling stone.

Evaluation of the superficial hardness is a useful method for distinguishing between these different types of lump. Even performing a crude test, by hand, using a needle point, allows these different types of lump to be effectively distinguished.

Following successful identification, all pieces of sand still attached to the surface of lumps must be removed using tools such as scalpels or needles. In order to remove as many pieces of sand as possible, this work should be done under the stereo microscope.

Great care should be taken in this phase as the sample is very delicate and easily damaged.

When the sample is clean, it must be weighed to check that the mass matches the requirements for the radiocarbon dating process (usually 20 mg). It can, then, be stored in a box with rigid walls to avoid damage during transit to the laboratory.

---

[6]This problem is particularly prevalent in masonry containing squared off blocks laid upon very thin mortar joints. In this case it is only possible to proceed if a cross section of the wall is accessible.

3.4 Treatment at the AMS laboratory

At the AMS laboratory the pure lime lumps are processed using the standard treatment for carbonate. Upon arrival the sample is analysed using an optical microscope to double check for the presence of macro contaminants. If necessary, further mechanical cleaning is carried out to remove these.

Once cleaned, samples are treated with hydrogen peroxide (H_2O_2) to remove the outside layer. Approximately half of the minimum amount of the pre-cleaned samples, by weight, is dried in an oven where it is also treated with H_2O_2 and converted to CO_2 using phosphoric acid (H_3PO_4).

The CO_2 extracted from the sample is, then, reduced to graphite using hydrogen (H_2) in the presence of an iron powder catalyst. The graphite is finally pressed into tablets which are used as a target in the accelerator mass spectrometer for measurement of the carbon isotopic ratios (Pesce et al., 2009; CEDAD, n.d.).

4. Applications

4.1 Literature review

In 2008 a detailed example of this sampling and dating method was presented during an international conference. This work involved the collection of two samples of lime lumps from the apses of the medieval church of S. Nicolò of Campodimonte (Camogli, Genoa - Italy) and dating with the radiocarbon method. Results obtained were evaluated and compared with the radiocarbon dating of organic material collected in the same apses. All the results were finally compared with results of other dating methods including mensiochronology of squared off blocks (Pesce et al., 2009).

Prior to this, relatively few researchers carried out radiocarbon dating of lime lumps in archaeological sites. Exceptions being Gallo, on the alto-Medieval castle of Aghinolfi (Massa Carrara - Italy; Gallo, 2001) and Fieni on the basilica of S. Lorenzo Maggiore in Milan (Italy; Fieni, 2002).

However, since 2008, additional tests have been made by the research group of the Institute for the History of Material Culture of Genoa (Italy). A total of 9 new samples taken from three different archaeological sites were dated following the procedure described earlier. Results were found to be consistent with the respective archaeological frameworks and uncertainty of radiocarbon dating was often reduced by comparison with other archaeological information. Among these samples, one was removed from the Medieval Castle of Zuccarello (Italy).

4.2 The Saint Nicolò of Capodimonte church

The San Nicolò church of Capodimonte (Camogli, Genoa – Italy; Fig. 2) is a medieval church on the Portofino Mountain with a crux commissa plan. During the most recent phase of restoration work a multidisciplinary approach was adopted to study its principal construction stages. The work also included archaeological analysis of the walls, research of bibliographic sources and radiocarbon dating of mortars.

Archaeological evidence and bibliographic sources suggested that the church existed in the 12th century, and underwent some major changes during the 12th and 13th centuries. Nevertheless, the lack of detailed documentation about the different construction phases suggested that a more direct and absolute dating method would be desirable.

Fig. 2. Medieval church of Saint Nicolò of Capodimonte (Camogli, Genoa - Italy; author Arch. Paola Cavaciocchi)

Radiocarbon dating was applied to two samples of lumps and a sample of charcoal. Sampling of lumps was possible through the large mortar joints (>1 cm thickness[7]) and damaged masonry. Samples were taken at a depth of 1-2 cm inside the wall to ensure that sufficient calcium carbonate was collected for the ^{14}C dating.

The first sample (sample n. 1 in tab. 1 and red mark in figg. 3 and 4) was removed from the wall next to the interior border of the apsidal basin, on the right-hand side of the transept. The second sample (sample 2 in tab. 1 and yellow mark in fig. 3 and 4) was removed from the external face of a wall, under the same apsidal basin. The sample of charcoal (sample 3 in tab. 1 and green mark) was removed from the internal face of the main apse.

Although the stratigraphic position of the first two samples was clear (sample n. 2 had to be older than sample n. 1), the position of the third sample (the charcoal) was not perfectly established as it could have been a part of the mortar used for the construction of wall, or a residual part of a more recent plaster applied on the interior face of the wall.

[7]This is a specific characteristic of the building technique of the oldest archaeological stages.

The selected samples were then submitted for ^{14}C dating at the Centre for Dating and Diagnostics at the University of Salento (Lecce, Italy). At the laboratory, the inorganic samples underwent the standard process used for carbonates whereas the charcoal sample was treated using the acid-alkali-acid protocol (Pesce et al., 2009). The 'before present' ^{14}C ages of the three samples are shown in tab. 1 together with the calibrated[8] and the expected ages.

Fig. 3. Sampling point inside the church (red mark: sample 1; yellow mark: sample 2; green mark: sample 3)

Sample 2 was ^{14}C dated to between the end of the 10th and the middle of the 12th century cal AD, sample 1 to the 11th–13th century cal AD, and the charcoal sample (n. 3) to the last construction phase of the building between the 12th and the 14th century cal AD.

Analysis of ^{14}C data shows that the ages obtained are consistent with each other, with their stratigraphic position (Fig. 4), and with the information obtained from historical sources. In fact, the archaeological analysis led to the identification of three main building phases, all dated to the Middle age. Bibliographic sources reveal that the church existed at least in the 12th century AD, when two congregations of monks undertook work on the building, while the last documented construction phase can be dated to the 15th century AD.

[8]These were obtained using OxCal v 4.0 software (Bronk Ramsey, 1995, Bronk Ramsey, 2001) and the IntCal04 atmospheric calibration curve (Reimer et al., 2004).

The first lump (sample n. 2) was selected from structures of unknown age but stratigraphically older than the 12th century building, while the second sample (sample n. 1) was selected from structures archaeologically dated to the 12th century. The charcoal (sample n. 3) has been ascribed to the last construction phase of the building dated to the 13th century AD based on the typology of the walls and on the sizes of stones[9].

Overall, the ^{14}C results obtained confirm the expected ages of the samples and their stratigraphic position.

Sample n.	Archaeological site	Laboratory number	^{14}C age (BP)	$\delta^{13}C$ (‰)	Calibrated age (AD)	Probability (%)	Expected age (century)
1	St. Nicolò of Capodimonte church	LTL2133A	917±55	-22,9±0,3	1010-1220	95,4	XII-XIII
2	St. Nicolò of Capodimonte church	LTL2978A	1005±45	-12,1±0,2	890-920	2,7	XI
					960-1160	92,7	
3	St. Nicolò of Capodimonte church	LTL2132A	790±60	-26,0±0,5	1040-1090	4,6	XIII-XIV
					1120-1300	90,8	
4	Zuccarello castle	LTL4756A	559±35	-4,8±0,3	1300-1370	49,9	XIII-XIV
					1380-1440	45,5	

Table 1. Radiocarbon ages obtained for samples 1 to 4

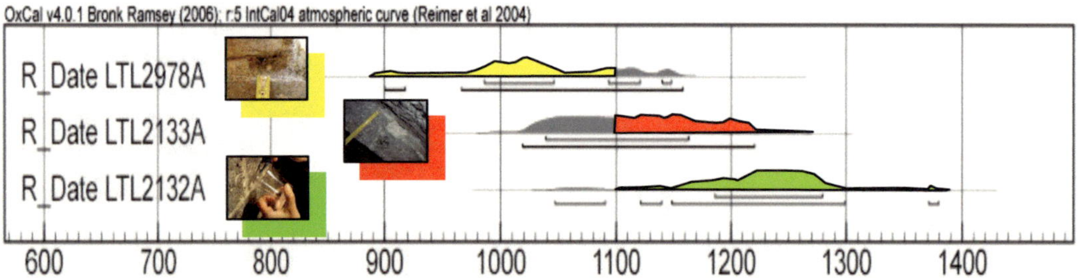

Fig. 4. St. Nicolò of Capodimonte church: comparison of the calibrated date. According to the previous pictures, red colour identifies the sample 1, yellow colour identifies the sample 2 and green colour identifies the sample 3. Coloured areas in the calibrated curve show the most likely period of time for the single result after comparison with other dating methods used in this archaeological analysis

4.3 Zuccarello castle

The Medieval castle of Zuccarello (Savona, Italy; Fig. 5) is a ruined construction built on the top of a hill in the west side of the Liguria region (North West of Italy). In recent years it has undergone restoration work and archaeological studies.

[9]This is a well established archaeological dating method used in this area of Italy.

In order to obtain an archaeological dating of the main walls of this building several techniques were used including: mensiochronology of bricks, chronotypological dating of door and window frames, historical and artistic dating and the radiocarbon dating of a single lime lump from one of the oldest parts of the surrounding wall (Fig. 6).

Fig. 5. The medieval Castle of Zuccarello (Savona, Italy)

Fig. 6. Sampling point of the lime lump

Uncalibrated ^{14}C and calibrated ages of samples, obtained using OxCal v 3.10 and the IntCal04 atmospheric calibration curve software, are reported in table 1 and figure 7. These results show that, even though the radiocarbon determination exhibits a normal distribution (red line in the left hand side of the figure), the curve for calibrated data is divided into two parts with very similar probabilities (49.9% and 45.5%) because of the shape of the calibration curve in this time range.

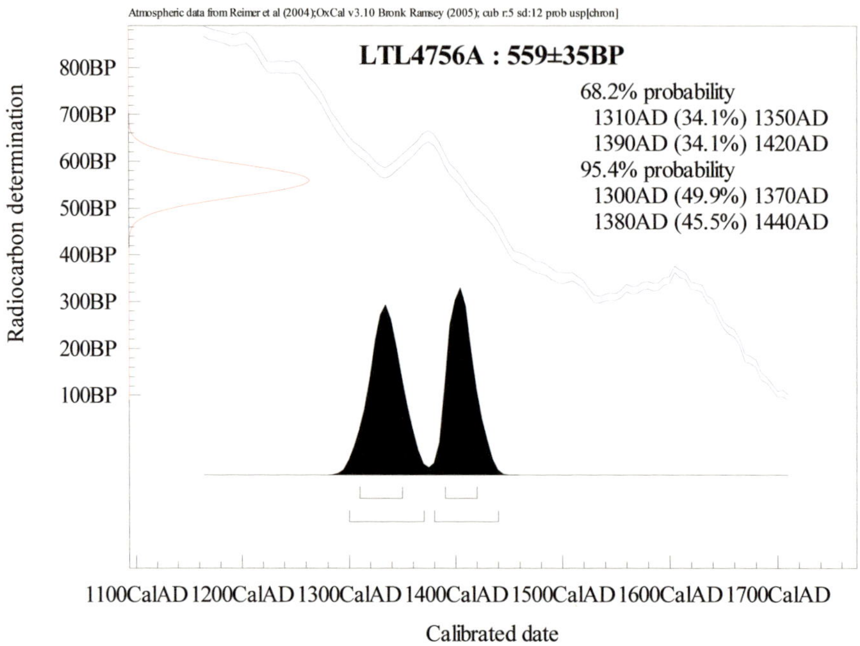

Fig. 7. Calibrated and uncalibrated age of the lime lump picked up in the lower part of the surrounding wall of the Zuccarello castle (analysis by CEDAD, Lecce – Italy)

The reliability and usefulness of this data was evaluated by comparison with results from other dating methods. Figure 8 shows collected dating estimates relative to a timeline plotted along the bottom of the figure. Each line in the figure corresponds to a specific dating obtained from different methods such as mensiochronology of bricks[10], chronotypology of doors or windows[11] and dating of artwork.

The lines corresponding to the mensiochronology method (black texture) were obtained from Five (M1-M5) groups of bricks collected in different parts of the castle. Their length represents the chronological range of production.

Above these lines the radiocarbon dating of the lime lump is reported (obliques lines on texture) and, above this, the results of the chronotypology dating (vertical lines on texture). Dating from the artistic evaluations of frescos, still visible inside the castle, are given at the top of the figure (obliques lines on texture).

[10]Mensiochronology of bricks is a well established archaeological dating method based on the trend of the size of bricks over the time. Its accuracy (a few tens of years) is not constant but varies over the time depending on various factors such as the amount of data available for a specific production defined in time and space.

[11]Chronotipology of door and window frames is a dating method based on the shape of frames of doors and windows with an accuracy of a few hundred years.

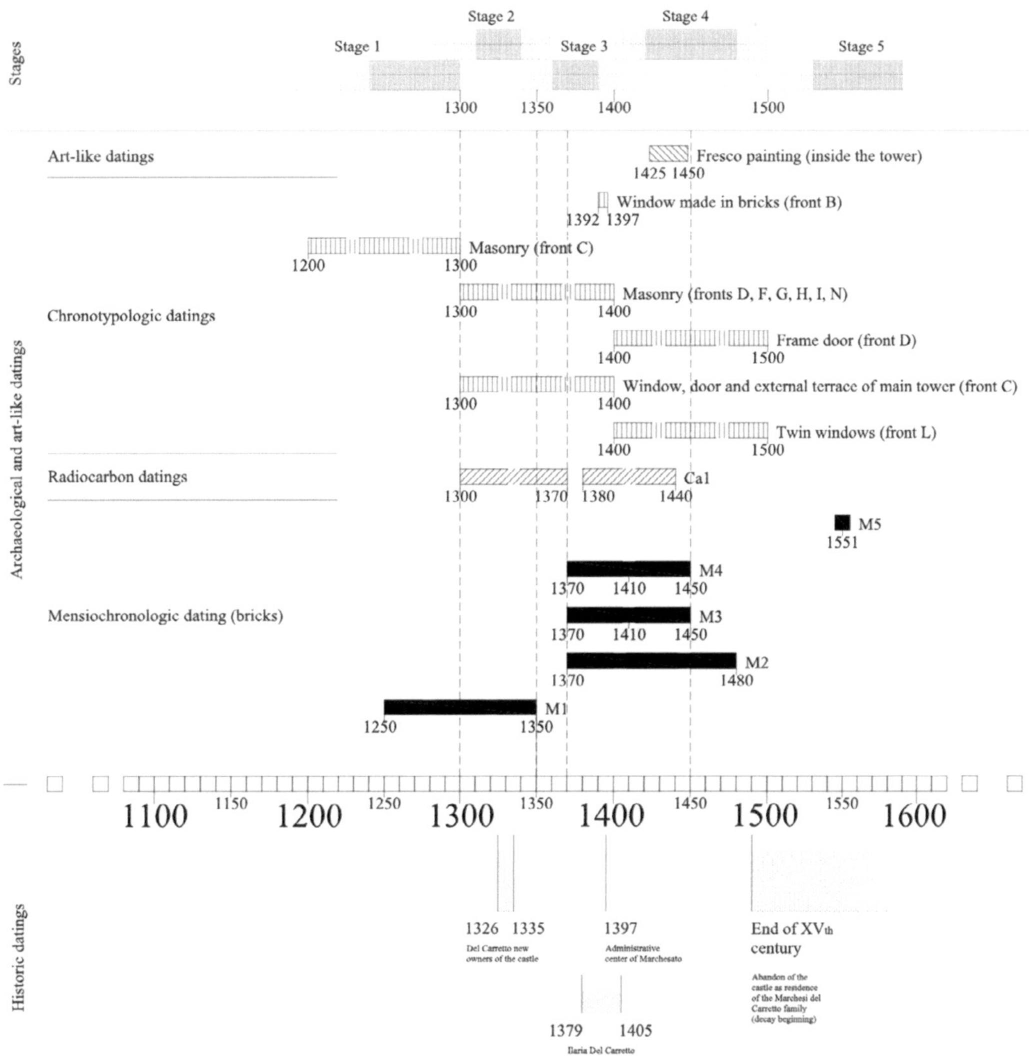

Fig. 8. Comparison of results obtained from all the different dating methods applied at Zuccarello castle

The graph highlights the agreement between radiocarbon dating and other dating methods. The first peak of the radiocarbon dating curve between 1300 and 1370 matches historical records indicating that between 1326 and 1335 the historically important family of "Del Carretto" acquired the castle (at the bottom of the figure). At this time significant expansion of the main building was carried out.

However, the second peak of the radiocarbon dating curve, between 1380 and 1440, is not in agreement with archaeological records describing building techniques of this period. In particular there are anomalies regarding the size, shape of masonry unit, mortar joints and the laying technique of the wall where the lump was sampled that does not fit the radiocarbon dating (whereas the first peak does).

Therefore, after comparing radiocarbon data with other archaeological evidence, the surrounding wall of the castle where the lime lump was picked up is believed to date from between 1300 and 1370.

5. Conclusions

Experience already gained through the studies described highlight the importance and future potential of radiocarbon dating of pure lime lumps for historic constructions.

Success of the method is dependent on following an appropriate sampling methodology which involves archaeological, chemical and physical knowledge of the building and the materials being studied. Following the removal of appropriate lime mortar samples from the historic site of interest the pure lime lumps must be extracted under the view of an optical microscope. Once extracted these lumps must be weighed and properly stored before being sent to the AMS laboratory.

Despite recent advances in this technique the full potential is still to be discovered. Unknowns include the dating of mixtures containing hydraulic lime which accounts for a significant proportion of historic mortar.

Re-carbonation of calcium carbonate is expected to affect the results of radiocarbon dating (Karkanas, 2007). However few studies have been carried out to address this phenomenon in lime lumps. This is believed due to the associated difficulties in identification of primary and secondary calcite. To date, results applied to lime lumps do not suggest this problem despite their extraction from ruined walls of archaeological sites where the penetration of rising damp or rain was expected has not been carried out.

6. Acknowledgment

The authors would like to thank Arch. Rosita Guastella, Arch. Paola Cavaciocchi and Arch. Carolina Lastrico for the study of the medieval church of S. Nicolò of Capodimonte. Thanks are extended to Arch. Simona Martini, Dr. Giorgio Casanova and Mauro Darchi for their job at Zuccarello Castle. Help is acknowledged from Dr. Gianluca Quarta and Prof. Lucio Calcagnile at the Centre for Dating and Diagnostics, University of Salento (Lecce, Italy) and to all colleagues of the Institute of the History of Material Culture of Genoa (Italy). Finally, the authors wish to dedicate this chapter to the memory of Professor Tiziano Mannoni who pioneered the archaeometric studies of historic mortars and has been a great influence and help.

7. References

Bakolas, A., Biscontin G., Moropoulou A. & Zendri E. (1995). Characterization of the lumps in the mortars of historic masonry. *Thermochimica Acta*, 269/270, pp. 809-816

Berger, R. (1992). ^{14}C dating mortar in Ireland. *Radiocarbon*, 34, 3, pp. 880-889

Bronk Ramsey, C. (1995). Radiocarbon calibration and analysis of stratigraphy: the OxCal program. *Radiocarbon*, 37, 2, pp. 425–30.

Bronk Ramsey, C. (2001). Development of the radiocarbon calibration program. *Radiocarbon*, 43, 2A, pp. 355–363

Bruni, S., Cariati, F., Fermo, P., Cariati, P., Alessandrini G. & Toniolo L. (1997). White lumps in fifth to seventeenth century AD mortars from northern Italy. *Archaeometry*, 39, 1, pp. 1-7

Bugini, R. & Toniolo, L. (1990). La presenza di grumi bianchi nelle malte antiche: ipotesi sull'origine. *Arkos Notizie GOR*, 12, pp. 4-8

CEDAD, n.d.. Sample preparation for AMS radiocarbon dating, In: *Internet site of CEDAD*, Date of access: November 2011, Available from: http://cedad.unisalento.it/en/clams.php

Elsen, J., Brutsaert, A., Deckers, M. & Brulet, R. (2004). Microscopical study of ancient mortars from Turnay (Belgium), *Materials Characterization*, 53, pp. 289-294

Elsen, J. (2006). Microscopy of historic mortars - a review. *Cement and Concrete Research*, 36, pp. 1416-1424

El-Turki, A., Ball, RJ. & Allen, GC. (2007). The influence of relative humidity on structural and chemical changes during carbonation of hydraulic lime. *Cement and Concrete Research*, 37, 8, pp. 1233-1240

Emrich. K., Ehhalt, DH. & Vogel, JC. (1970). Carbon isotope fractionation during precipitation of calcium carbonate. *Earth and Planetary Science Letters*, 8, 5, pp.363-371

Fieni, L. (2002). La Basilica di San Lorenzo Maggiore a Milano tra età tardoantica e medioevo: metodologie di indagine archeometrica per lo studio dell'elevato. *Archeologia dell'Architettura*, 7, pp. 51-98

Folk, R.L. & Valastro, S. Jr. (1976). Successful technique for dating of lime mortar by carbon-14. *Journal of Field Archaeology*, 3, pp. 203-208

Franzini, M., Leoni, L., Lezzerini, M. & Sartori, F. (1990). On the binder of some ancient mortars. *Mineralogy and Petrology*, 67, pp. 59-69

Gallo, N., Fieni, L., Martini, M., Sibilia, E. (1998). Archèologie du bati, ^{14}C et thermolumiscence: deux exemples en comparaison, *Actes du 3eme Congrés International 14C et Archéologie*, Lyon, 6-10 Avril

Gallo, N. (2001). ^{14}C e archeologia del costruito: il caso di Castello Aghinolfi (MS). *Archeologia dell'architettura*, 4, pp. 31-36

Gliozzo, E. & Memmi Turbanti, I. (2006). La catodoluminescenza e l'analisi di immagine per lo studio delle malte. *Proceedings of the IV Congresso Nazionale di Archeometria - Scienza e Beni Culturali*. Pise, Febraury 2006.

Goren, Y. & Goring-Morris, AN. (2008). Early Pyrotechnology in the Near East: Experimental Lime-Plaster Production at the Pre-Pottery Neolithic B Site of Kfar HaHoresh, Israel. *Geoarchaeology: An International Journal*, 23, 6, pp. 779-798

Hale, J., Heinemeier, J., Lancaster, L., Lindroos, A., & Ringbom, A. (2003). Dating ancient mortar. *American Scientist*, 91, pp. 130-137

Ingham, JP. (2005). Investigation of traditional lime mortars - the role of optical microscopy. *Proceedings of the 10th Euroseminar on Microscopy Applied to Building Materials*, University of Paisley, June 2005, pp. 1-18

Karkanas, P. (2007). Identification of lime plaster in prehistory using petrographic methods: a review and reconsideration of the data on the basis of experimental and case studies. *Geoarchaeology: An international journal*, 22, 7, pp. 775-796

Leslie, AB & Hughes, JJ. (2002). Binder microstructure in lime mortars: implications for the interpretation of analysis results. *Quarterly Journal of Engineering Geology and Hydrogeology*, 35, pp. 257-263

Marzaioli, F., Lubritto, C., Nonni, S., Passariello, S., Capano, M. & Terrasi, F. (2011). Mortar Radiocarbon Dating: Preliminary Accuracy Evaluation of a Novel Methodology. *Analytical Chemistry*, 83, 6, pp. 2038-2045

Morse, J.W. & Meckenzie, F.T. (1990). *Geochemistry of Sedimentary Carbonates* (1st edition), Elsevier, 0-444-88781-4, Amsterdam

Nawrocka, D., Michniewicz, J., Pawlyta, J., & Pazdue, A. (2005). Application of radiocarbon method for dating of lime mortars. *Geochronometria*, 24, pp. 109-115

Norman, D. (1961). *A history of building materials*, Phoenix House, London

Oates, JAH. (1998). *Lime and limestone: chemistry and technology, production and uses*, Wiley-VCH, 3-527-29527-5, Weinheim

Pachiaudi, C., Marechal, J., Van Strydonck, M., Dupas, M., & Dauchot-Dehon, M. (1986). Isotopic fractionation of carbon during CO_2 absorption by mortar. *Radiocarbon*, 28, 2A, pp. 691-697

Pesce, G.L.A., Quarta, G., Calcagnile, L., D'Elia, M., Cavaciocchi, P., Lastrico, C. & Guastella, R. (2009). Radiocarbon dating of lumps from aerial lime mortars and plasters: methodological issues and results from the S. Nicolò of Capodimonte Church (Camogli, Genoa - Italy). *Radiocarbon*, 51, 2, pp. 867-872

Pesce, G.L.A. (2010). Radiocarbon dating of lumps of no completely mixed lime in old constructions: the sampling problem, *Proceedings of 2nd Historic Mortar Conference*, Prague, September 2010

Pesce, G.L.A., Ball, R., Quarta, G., & Calcagnile, L. (2012). Identification, extraction and preparation of reliable lime samples for the ^{14}C dating of plasters and mortars with the method of "pure lime lumps", *Proceeding of the 6th International Symposium on Radiocarbon and Archaeometry*, Cyprus, April 2011

Rattazzi, A. (2007), *Conosci il grassello di calce? Origine, produzione, impiego del grassello di calce in architettura, nell'arte e nel restauro*, Edicom Edizioni, 978-88-86729-70-3, Monfalcone

Reimer, PJ., Baillie, MGL., Bard, E., Bayliss, A., Beck, JW., Bertrand, CJH., Blackwell, PG., Buck, CE., Burr, GS., Cutler, KB., Damon, PE., Edwards, RL., Fairbanks, RG., Friedrich, M., Guilderson, TP., Hogg, AG., Hughen, KA., Kromer, B., McCormac, G., Manning, S., Bronk Ramsey, C., Reimer, RW., Remmele, S., Southon, JR., Stuiver, M., Talamo, S., Taylor, FW., Van der Plicht, J. & Weyhenmeyer, CE. (2004). IntCal04 terrestrial radiocarbon age calibration, 0–26 cal kyr BP. *Radiocarbon*, 46, 3, pp. 1029–58

Sonninen, E. & Junger, H. (2001). An improvement in preparation of mortar for radiocarbon dating. *Radiocarbon*, 43, 2A, pp. 271-273

Turner, JV. (1982). Kinetik fractionation of carbon-13 during calcium carbonate precipitation. *Geochimica et Cosmochimica Acta*, 46, 7, pp. 1183-1191

Van Balen, K. (2005). Carbonation reaction of lime kinetics at ambient temperature. *Cement and Concrete Research*, 35(4), pp.647-657

Van Strydonck, M., Dupas, M., Dauchot-Dehon, M., Pachiaudi, Ch. & Maréchal, J. (1986). The influence of contaminating (fossil) carbonate and the variations of $\delta^{13}C$ in mortar dating. *Radiocarbon*, 28, 2A, pp. 702–10

Van Strydonck, M.K., Van der Borg, A.F.M., & De Jong, E.K. (1992). Radiocarbon dating of lime fractions and organic material from buildings. *Radiocarbon*, 34, 3, pp. 873-879

Vecchiattini, R. (2010). *La civiltà della calce. Storia, scienza e restauro*, De Ferrari, 978-88-6405-089-8, Genoa

Section 2

Applications

3

Holocene Soil Chronofunctions, Luochuan, Chinese Loess Plateau

Gang Liu, Wennian Xu, Qiong Zhang and Zhenyao Xia
China Three Gorges University
China

1. Introduction

Since soil genesis can hardly be observed directly over decades and centuries, the research of soil chronosequences is the most suitable way to assess quantitative knowledge on soil development (Bockheim, 1980). A soil chronosequence is a quantitative description of how the soils properties in a given area change with time (Vincent, 1994). Soil chronofunctions have been equated with the mathematical expression of chronosequence data, typically utilizing correlations and curve-fitting, or some derived combination. They are useful for studying pedogenesis, relative dating of surfaces and geologic events, and for predicting recovery rates of disturbed soils.

Therefore, numerous studies on chronosequence investigation and chronofunction construction were carried out in the last decades. The classic chronosequences were established on moraines left in the wake of retreating glaciers (Mellor, 1987; Mattews, 1992) and on coastal sand dune systems (Olson, 1958). More recent work on historical chronosequences was investigated on a much wider range of landscapes. In the Luquillo experimental forest, Puerto Rico, a soil chronosequence formed in landslide scars showed that the base saturation index and nutrient cation pools in surface mineral soil increases during primary succession (Zarin and Johnson, 1995). Carreira et al. (1994) evaluated total and mineral N concentration, net nitrogen mineralization, and nitrification in soils from a burnt patches chronosequence in southeast Spain. The results indicated that an increasing fire frequency in the last few decades was associated with a sharp decrease in surface soil organic matter and total nitrogen concentrations and pools, and with changes in the long-term nitrogen dynamic patterns. In the western Brazilian Amazon Basin state of Rondonia, the net nitrogen mineralization and net nitrification in soil profiles along a tropical forest-to-pasture chronosequence were examined to investigate possible mechanisms for changes to soil nitrogen sources and transformations that occur as a result of land use (Piccolo et al., 1994). Marine terraces were also suitable for constructing soil chronosequences. Merritts et al. (1991) evaluated six pedogenic properties of soils from chronosequences developed on two flights of unlifted marine terraces in northern California. They found that the soil properties have systematic and time-dependent trends, but each property differs in the rate of change. An example of chronosequences produced on old mining areas is a study on development rate of soil from abandoned phosphate-mined sites on Nauru Island, Central Pacific (Manner and Morrison, 1991).

Most of the soil chronosequence studies suggest linear, power, exponential or logarithmic changes of soil properties with time. Bockheim (1980) used three linear and non-linear models to construct chronofunctions for fifteen soil properties of thirty-two chronosequences from seven types of parent materials, including till, aeolian sand, alluvium, mine spoil, volcanic ash, raised beach deposits, and mudflow. The single-logarithmic model showed the best relationship between soil property and time, and more than 85% of the correlation coefficients were statistically significant. Birkeland (1984) investigated two soil chronosequences from the high country of the South Island of New Zealand to establish chronofunction. Out of a possible twenty one chronofunctions, sixteen were significant at the 0.05 level. Most of the chronofunctions best fit a power model, and the remaining ones are split between linear and exponential models. Chemical and mineralogical characteristics have been determined for a chronosequence of six soil profiles in the western Cairngorms of Scotland (Bain et al., 1993). Exchangeable Ca and Mg decrease with time and base saturation decreases exponentially from 24.6% in the Ah horizon of the youngest profile to 2.8% in the comparable horizon of the 10,000 year old profile according to a exponential function. A soil chronosequence in the northern shore of Lake Michigan, USA was investigated (Barrett, 2001). The results indicated that for most forms of extractable Fe and Al, both linear and log-linear functions adequately describe the changes in property with surface age.

Although linear, power, exponential or logarithmic functions are widely used, they are not always the best option for chronofunctions. Polynomial, hyperbolic, S-shaped, sigmoidal, logistic curve or two separate linear regression lines might improve not only the fit of the chronofunction but also advance our understanding of the pedologic system. Jacobson and Birks (1980) showed a curve in which loss on ignition of A horizons in Canadian soils follows an approximate sigmoidal relationship when plotted against time. For some soil properties, rates of pedogenesis are initially very rapid and then decrease with time, quickly approaching an upper asymptote. This type of data might be easily fitted to hyperbolic functions (Mellor, 1985). Polynomial functions have been successfully used to fit the time and soil properties in soil chronosequences at three sites within the Transantarctic Mountains (Bockheim, 1990). Selection of the appropriate chronofunction model should be both reasonable with respect to fit, as well as theoretically matching the expected temporal trend (Schaetzl et al., 1994).

The building and interpreting of soil chronosequences are hampered by several well-known problems. A major difficulty is dating of soil to establish a precise and reliable time scale. In last decades, many Quaternary dating methods have been improved and developed, including rock stratigraphy, magnetic stratigraphy, archaeology, Ar-Ar laser microprobe dating, Thermal Ionization Mass Spectrometer, Thermol Uminescence dating, Optic Stimulated Luminescence, Electron Spin Resonance, and other isotope dating method, etc. For the loess, two dating methods are always used. One is Thermoluminescence dating which has been applied to loess from the last 100-200 ka. This method is based on several assumptions. During transport, dust is exposed to sunlight, which releases the energy from the crystals and, therefore, resets the clock. After deposition, the loess is radiated by cosmic rays and the primordial radioactive isotopes of U, Th and K and their daughters. Another important dating method based on radioactive decay is ^{14}C dating. The radiocarbon dating has added precision to existing knowledge regarding the many feet of unconsolidated loessial materials that were deposited alternately during a few thousand years of Holocene time in response to environmental fluctuations (Miao, 2005; Muhs et al., 2004). However, due to the relatively short half-life of 5730 yr, it is applicable only to the last ca. 40 ka. ^{14}C dating also requires well-defined carbon material of organic origin to guarantee that it reflects the atmospheric $^{14}C/^{12}C$ ratio. Only material that has never exchanged with other carbon (e.g., $CaCO_3$) present in the loess should be used. Finally, one must be aware that ^{14}C ages have to be calibrated, and at present, the necessary calibration curve covers only the last 13 ka.

Another difficulty is holding all soil-forming factors constant except time. In particular, climate is highly unlikely to have held steady, even for short periods. Many studies have been implemented to research paleoclimatic change on Chinese Loess Plateau. Micromorphological investigations on loess-paleosol sequences in Potou section near Luochuan showed that the paleosols with an Ah-C horizon sequence are genetically comparable to the natural Holocene soil and are most probably formed under the environmental conditions of steppe vegetation. The soils with a Bt-Ck horizon sequence - the Ah-horizon is always eroded - are forest soils, developed under moister conditions than the Holocene climate (Bronger and Heinkele, 1989). Based on chemical (major and trace element) and isotopic (Sr, Nd) analyses of the Luochuan loess-paleosol sequence, Gallet et al. (1996) concluded that systematic variations of element abundances and ratios between loess and paleosols can be used as chemical indicators for pedogenetic intensity and so for paleoclimatic change. But the correlation between precipitation and chemistry was not exist in Jahn et al.'s (2001) study in other three sites (Xining, Xifeng, Jixian) on the Chinese Loess Plateau. Maher et al. (1994) proposed a mineral magnetic approach for estimating palaeoprecipitation across the Chinese Loess Plateau. The correlation between pedogenic magnetic susceptibility and contemporary annual rainfall was used to reconstruct rainfall for interglacial and glacial episodes spanning the entire Quaternary. It was concluded that for the last interglacial, the presently dry western sites received up to 60% more rainfall per year (compared to present), and the presently humid south and east sites up to 30% more. ^{14}C research of Late Pleistocene and Holocene fossil soils and cultural horizons of archaeological sites also helps reconstructing and modelling natural and climatic changes (Morozova, 1990).

Despite the problems involved in their construction, there is no doubt that soil chronosequences are immensely powerful tools for probing the rate and direction of soil evolution. Indeed, they are the only way of establishing how pedogenesis operates over centuries and longer periods. Well-dated chronosequences are therefore a boon to pedologists. They are also invaluable to geomorphologists, for, once a soil chronosequence is established, it may be used to investigate other landscape processes. After a chronofunction has been obtained from a data set, the y-intercept may also be used to reconstruct the situation at time zero (Schaetzl et al., 1994).

The perfectly preserved Heilu soil could provide important information about environmental evolvement on Chinese Loess Plateau in the Holocene (Tang & He, 2004; Liu, 2009). Heilu soils are principal soil components on gently rolling or undulating land surfaces, of the Loess Plateau where past erosion has been limited. Several studies on chronosequence of Heilu soil profile were carried out, but the chronofunctions were seldom descibed (Hu, 1994; Chen et al., 1998; Tang & He, 2002). The aim of this study was to characterize physical-chemical properties in a well-constrained and dated Holocene Heilu soil profile, establish soil chronofunctions, in order to improve the understanding of soil formation and development with time, to provide theoretical basis for prediction of soil recover and data for modeling soil genesis (Finke & Hutson, 2008).

2. Materials and methods

2.1 Study area

The Loess Plateau in northwest China covers an area of 530,000 km^2, larger than Spain and almost as large as France; the loess cover largely ranges in thickness from 30 to 80 m. The loessial soils are characterized by its yellowish colors, absence of bedings, silty texture, looseness, macroporosity and collapsibility when saturated. The Loess Plateau is conveniently divided into three zones: sandy loess in the northern part, typical loess in the middle, and clayey loess in the south (Liu, 2009). The soil chronosequence was located in Luochuan (N35°42.561', E109°23.952'), where the landform is a loess tableland, in the middle of Chinese Loess Plateau (Figure 1). The climatic conditions of this area are characterized by warm and semi-humid continental monsoon climate. Mean annual precipitation in Luochuan is about 622 mm, concentrated in the summer (July to September). Conversely, the climate is cold and dry in winter, with low rainfall. The mean annual temperature range is 9.2°C. The present-day vegetation in the study area is dominated by grasses and shrubs (*Poaceae*, *Leguminosae*, *Labiatae*, *Rhamnaceae*, and *Compositae*).

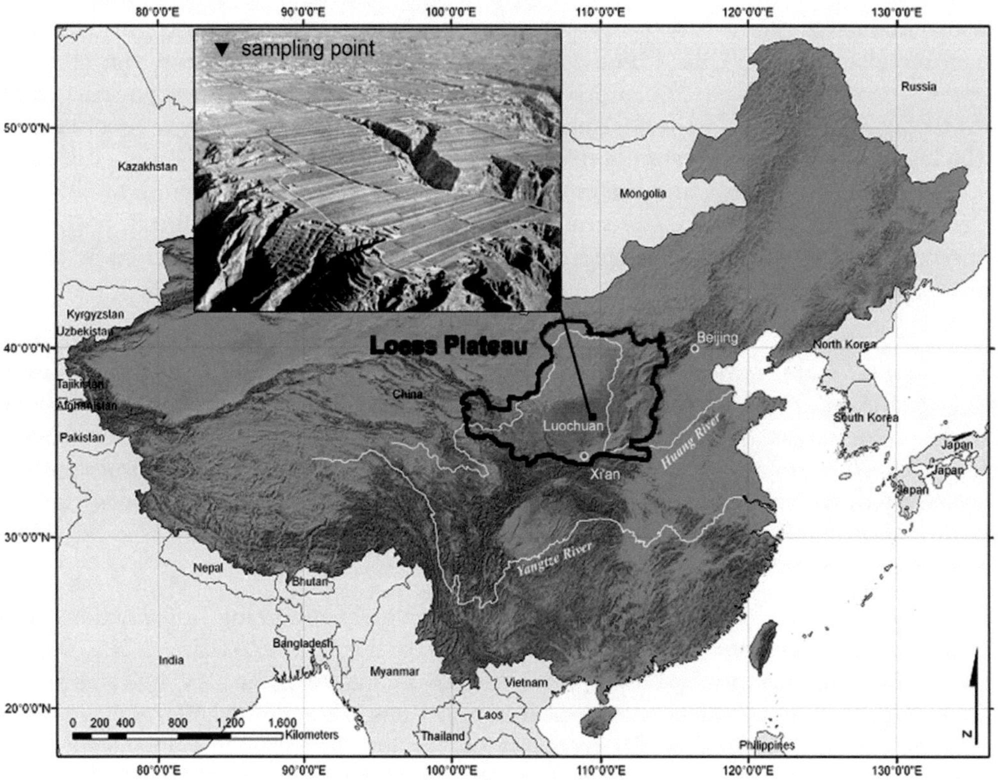

Fig. 1. Location of the soil chronosequence and relief of the study area at Luochuan, China.

2.2 Field work and laboratory analyses

The soil layers were sampled from the face of a hand-dug pit, approximately 50 cm×200 cm at the top and about 200 cm deep. Standard morphological description was made and the samples were uniformly scraped from the face of pit at 10-cm intervals. The interval was occasionally modified to avoid sampling across a horizon boundary.

In the laboratory, after air-drying and picking out visible roots and stone fragments, samples were passed through a 2-mm sieve. The amounts of organic carbon and carbonate-carbon were determined by a carbon analyzer with stepped heating routine (LECO RC 412), successively measuring both carbon fractions in two replicates. The soil samples were treated with H_2O_2 and HCl to remove organic matter, prepared with $Na_4P_2O_7$ and treated with ultrasound for 3 min before analysis. A standard procedure was used for operating the Malvern Mastersizer 2000. The pH was determined in deionised water using a soil:water ratio of 1:2.5. The total elements, including Mn, Fe, K, Na, Ca, Mg, and Zr were determined by X-ray fluorescence analysis (XRF) of fused discs. P was extracted sequentially by the procedure developed by Hedley et al. (1982) and measured by an absorption spectrophotometer.

For all soil samples, the extraction of humin and radiocarbon dating were carried out at the AMS ^{14}C laboratory of Institute of Earth Environment, Chinese Academy of Sciences, China. All soil samples were oven dried (60°C) to constant weight, then ground and sieved. In order to extract humic acids and date the humin fraction only, samples are commonly treated with acid and alkali (Mook and Streurman, 1983). All pre-treated samples were combusted by sealing the sample with CuO wire in an evacuated quartz tube, then placing the tube in a 950°C oven for 2 hours. The resulting CO_2 was purified and reduced to graphite targets for AMS using the apparatus and methods described in Vogel et al. (1987). The calibration of the ^{14}C age used the methods of Bronk Ramsey (2001) and Reimer et al. (2004).

2.3 Data evaluation

The results of the laboratory analyses were evaluated by analyzing soil depth functions and chronofunctions. In this study, three kinds of usually considered functions (Bockheim 1980, Schaetzl et al. 1994) including linear ($Y = a + bX$), logarithmic ($Y = a + b\ lnX$), and third order polynomial ($Y = a + bX + cX^2 + dX^3$) were used to fit the relationships between soil properties (Y) and soil ages (X). We selected the appropriate chronofunction model by the coefficient of determination (r^2), as well as theoretically matching the expected temporal trend.

3. Results and discussion

3.1 Soil morphology and classification

Zhu (1983) proposed the name Heilu for a kind of soil which differs appreciably from the regional light "Shestnut" soils and "Kastanozems". Heilu soils have brownish dark grey humus horizons at the surface that are usually 60-100 cm thick. Below the humus-rich horizons, deeper horizon is light gray-brown in color. Predominant textures in the profile are light to medium loam although the humus-rich A horizon showed from 8.5% to 22% clay. Illuvial Carbonates precipitate of deeper horizon appear chiefly in pseudomycelial and powdery forms. Small numbers of carbonate microconcretions are also present with depth, both in the soil and the underlying loess material. According to all morphological traits, the sampled pedon was classified as Heilu type soil in this study (Table 1). The ages of the Heilu soil profile was ranging from 841±32 to 12816±40 radiocarbon years BP (Figure 2). Figure 2 illustrated good linear relationships between soil age and depth with a high coefficient of determination (R^2). The slope of function showed average loess formation rate of profile in the Holocene was 0.0174 cm/a.

Horizon	Depth (cm)	Description
Ap	0~21	Light grayish brown, 10YR7/6 (moist), clay loam, weak medium granular structure, slightly hard (dry), abundant roots, some carbonate pseudomycelia in small spots, gradual smooth boundary.
Ah	21~80	Darkish brown, 7.5YR3/3 (moist), clay loam, fine subangular blocky structure, hard (dry), abundant roots, some pseudomycelia in root channels, a few krotowinas filled with C-material, smooth boundary.
AB	80~100	Color heterogeneity, darkish brown blocks in pale brown soils, clay loam, fine subangular blocks break to granules, slightly hard (dry), few roots, plenty of pseudomyceli, smooth boundary.
Bk	100~129	Very pale brown, 7.5YR5/3 (moist), clay loam, massive, hard (dry), some very fine pores, few pseudomycelia, some lime nodules of 0.5-1 cm in diameter, few roots, clear boundary.
C	129~200	Yellowish brown, 10YR8/4 (moist), clay loam, blocks break to granules, slightly hard (dry), few roots.

Table 1. Soil horizon designations

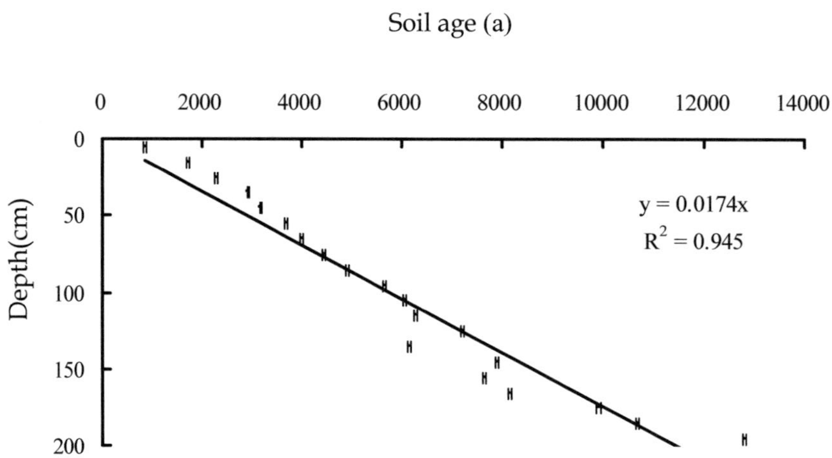

Fig. 2. The linear relationship between radiocarbon age and depth in the profiles. x and y in the equations is soil age and depth, R^2 is coefficient of determination. Bars represent ± error of measurement.

The most noticeable pedogenic processes in the field are organic carbon accumulation and calcium carbonate leaching in the upper centimeters of the soils, and illuviation further down the profile a few decimeters below. The calcium carbonate accumulations are very intensive in the Holocene soils.

3.2 Particle size distribution

The particle size distribution clearly reflected the predominance of silt (0.002-0.02 mm) and clay (<0.002 mm) (by International Particle Size Classification System) in the Ah horizon, where the soil age was about between 1500 and 4200 a B.P., indicating that clay formation was mainly restricted to the upper layer. Fining of upper horizons with increasing age has often been reported in chronosequences (Barrett, 2001).

Logarithmic functions for different particle classification were not statistically significant (Table 2). The corresponding linear functions were statistically significant except silt. However, all of them were rejected due to relatively low coefficients of determination. Third order polynomial models for three particle classifications were statistically very significant, and with higher coefficients of determination.

Equations	Classification	A	b	c	d	r^2
$Y = a + bX$	Clay*	17.39	-2.00×10^{-4}			0.26
	Silt	42.05	-4.00×10^{-4}			0.16
	Sand*	40.57	6.00×10^{-4}			0.20
$Y = a + b\ lnX$	Clay	21.20	-0.62			0.07
	Silt	42.86	-0.35			0.01
	Sand	35.95	0.97			0.02
$Y = a + bX + cX^2 + dX^3$	Clay**	12.88	24.00×10^{-4}	-4.00×10^{-7}	2.00×10^{-11}	0.59
	Silt**	31.38	53.00×10^{-4}	-8.00×10^{-7}	3.00×10^{-11}	0.79
	Sand**	55.74	77.00×10^{-4}	1.00×10^{-6}	-5.00×10^{-11}	0.75
*α<0.05; **α<0.01						

Table 2. Regression equations for different particle-size classification

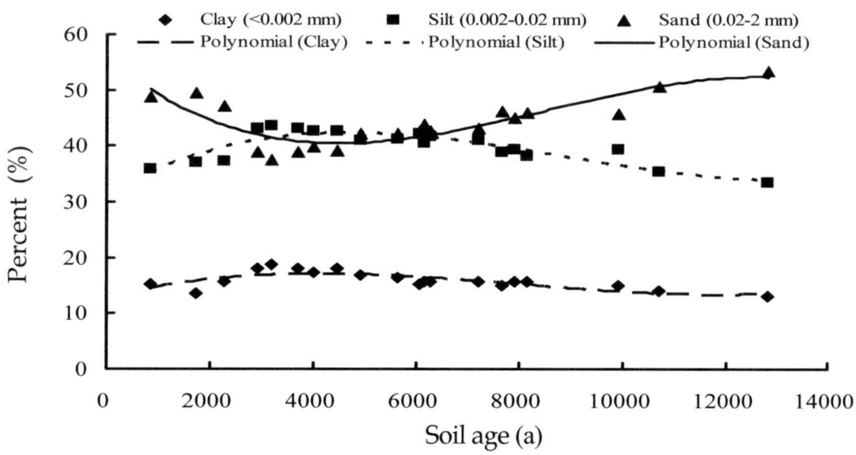

Fig. 3. Different particle-size classification as a function of soil age and regression curves of three order polynomial function

Our results contrast with Bockheim (1980) and Merritts et al. (1991) who found that logarithmic functions could adequately describe the relationship between clay and soil age, and Barrett (2001) who reported that changing silt content with time could be best fitted by linear models. The difference may be attributed to appearance of a humus horizon (Ah) which had greeter contents of clay and silt, and less sand than Ap and AB horizons (Figure 3). The clay in Ah horizon of the Heilu soil was derived from the in situ weathering of sand-sized particles, but not from illuviation from upper Ap horizon. The dry climate in this area conditioned a lower weathering degree and clay formation in the Ah horizon.

3.3 Organic carbon, calcium carbonate and pH

Linear, logarithmic and third order polynomial functions for organic carbon were statistically very significant and with very high coefficients of determination (around 0.8) (Table 3). But the intercepts of linear and third order polynomial functions, 0.57 and 0.6, were lower than 0.66 in surface layer. Theoretically, the organic carbon content in surface soil should be higher than any underlying soil layer (Chen, 2005). Data from Tang & He (2004) studies on Heilu soil agreed with our results from in Luochuan. Therefore, the intercept of logarithmic function, 1.68, was suitable, and it was the best fitting chronofunction for organic carbon content (Figure 3).

Equations	Property	*a*	*b*	C	*d*	r^2
$Y = a + bX$	Organic carbon**	0.57	-3.00×10^{-5}			0.79
	Calcium carbonate**	-2.87	16.00×10^{-4}			0.74
	pH**	8.10	2.00×10^{-5}			0.80
$Y = a + b\,lnX$	Organic carbon**	1.68	-0.15			0.79
	Calcium carbonate**	-51.39	6.83			0.59
	pH**	7.48	0.09			0.90
$Y = a + bX + cX^2 + dX^3$	Organic carbon**	0.60	-5.00×10^{-5}	8.00×10^{-10}	2.00×10^{-14}	0.81
	Calcium carbonate**	4.78	-41.00×10^{-4}	1.00×10^{-6}	-5.00×10^{-11}	0.85
	pH**	8.02	6.00×10^{-5}	-7.00×10^{-9}	3.00×10^{-13}	0.88
**α<0.01						

Table 3. Regression equations for organic carbon, calcium carbonate and pH

Though both linear and logarithmic functions for calcium carbonate were statistically very significant and have relatively high r^2 (0.74 and 0.59). Their intercepts were negative (-2.87 and -51.39) (Table 3). Hence, they can not describe the relationship between calcium carbonate content and soil age. However, this relationship could be best fitted by a third order polynomial function which has high r^2 value (0.85). This chronofunction showed redistribution of calcium carbonate, dissolution in the upper horizon, followed by reprecipitation and enrichment in the lower horizon, and the lowest value was observed in Ah horizon (Figure 4). This trend agrees well with the results of Tang & He (2004) and Zhao et al. (2006) for Heilu soil.

All of three functions for pH were statistically very significant and had high r^2 values (Table 3), indicating that theory must be utilized and incorporated when selecting equations. Huang & Gong (2001) and Jia et al. (2004) observed that pH values increased with soil age rapidly due to alkalinization which was caused by leaching at initial stage of soil formation, and the increase in lower horizons was slower. The logarithmic function for pH could show this trend best (Figure 4).

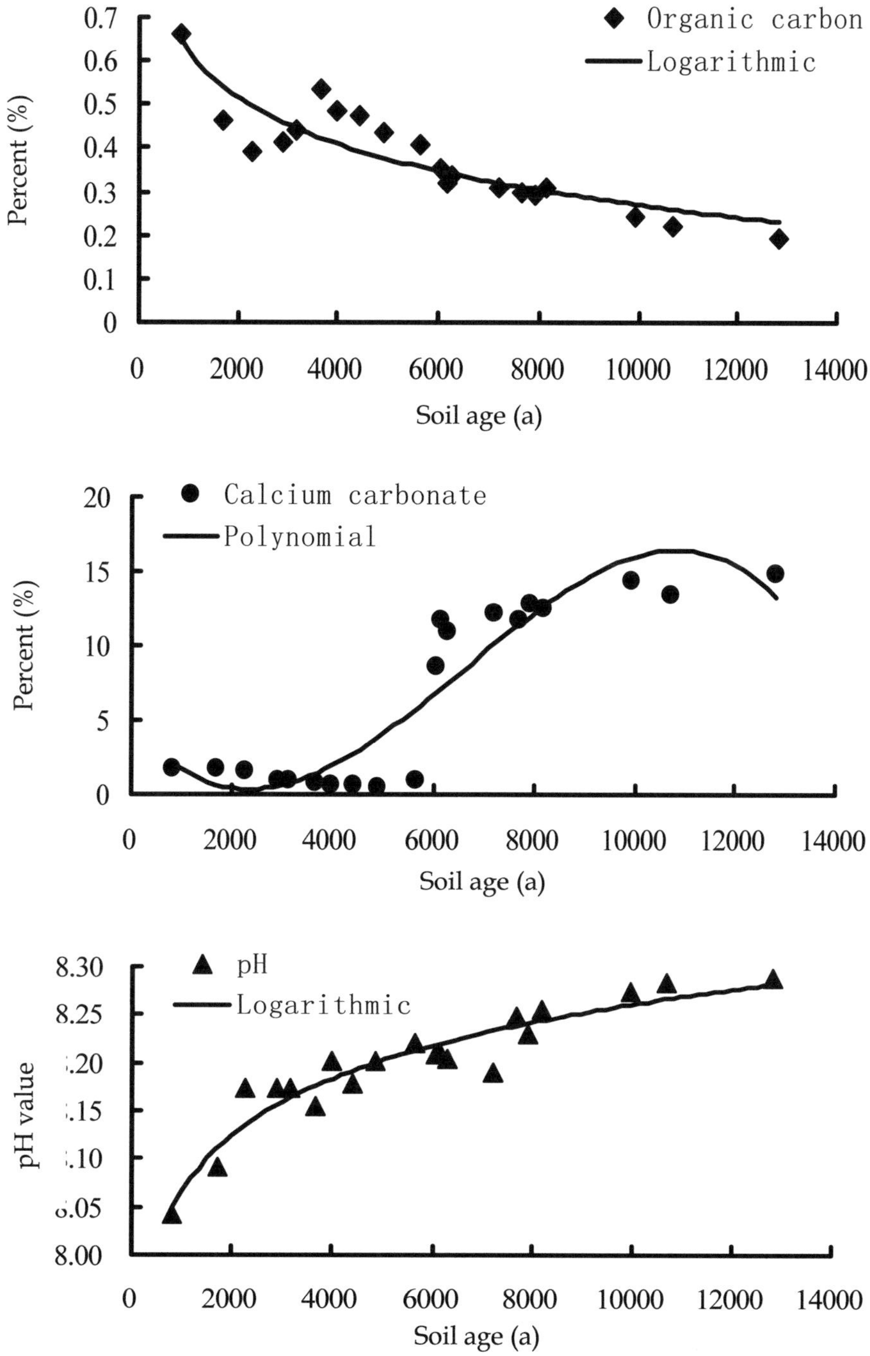

Fig. 4. The content of soil organic carbon, calcium carbonate, and pH as a function of soil age and regression curves

3.4 Mobility of elements in soil

In many studies, of the relations between selected mobile elements and Ti or Zr were used to identify enrichment and depletion of elements with time (Egli & Fitze, 2000). Since Ti can also be mobilized, Zr is considered more suitable for this purpose (Langley-Turnbaugh & Bockheim, 1998; Sauer et al., 2007).

Logarithmic and third order polynomial functions for Mn/Zr, Fe/Zr and K/Zr were statistically very significant (Table 4). They showed similar trends, in which ratio increased sharply before about 3000 a B.P., being constant in the older soils (Figure 5). It indicated that mobility or removal of Mn, Fe and K with time was greater in the upper centimeters, whereas depletion or enrichment were negligible in the older lower layers. The third order polynomial functions were more fitted than logarithmic functions in this study, due to higher r^2 (Table 4). On the other hand, the relationships between Mn, Fe, K and time were best fitted by linear and logarithmic functions in studies elsewhere (Lichter, 1998; Egli & Fitze, 2000; Sauer et al., 2007), probably attributed to different pedological contexts. These results agree well with Tang & He (2004).

Equations	Ratio	*a*	*b*	*c*	*d*	r^2
	Mn/Zr*	217.22	3.00×10^{-3}			0.22
	Fe/Zr**	10362.00	0.22			0.44
	K/Zr**	6175.20	0.15			0.68
$Y = a + bX$	Na/Zr**	4092.90	0.10			0.62
	Ca/Zr**	-4439.10	2.18			0.77
	Mg/Zr**	3527.00	0.27			0.92
	P/Zr**	141.63	0.01			0.68
	Mn/Zr**	61.71	20.39			0.46
	Fe/Zr**	809.08	1276.10			0.67
	K/Zr**	660.09	751.88			0.78
$Y = a + b\ lnX$	Na/Zr**	1358.60	392.23			0.41
	Ca/Zr**	-68638.00	9049.10			0.61
	Mg/Zr**	-5071.10	1197.70			0.82
	P/Zr**	-181.56	44.99			0.61
	Mn/Zr**	153.32	0.04	-5.00×10^{-6}	2.00×10^{-10}	0.69
	Fe/Zr**	7664.10	1.65	-2.00×10^{-4}	7.00×10^{-9}	0.78
	K/Zr**	5455.40	0.48	-4.00×10^{-5}	9.00×10^{-10}	0.81
$Y = a + bX + cX^2 + dX^3$	Na/Zr**	4894.30	-0.44	9.00×10^{-5}	-4.00×10^{-9}	0.79
	Ca/Zr**	5832.50	-5.34	14.00×10^{-4}	-7.00×10^{-8}	0.88
	Mg/Zr**	3811.10	-0.04	7.00×10^{-5}	-4.00×10^{-9}	0.96
	P/Zr**	165.90	0.02	6.00×10^{-6}	-3.00×10^{-10}	0.84
*α<0.05; **α<0.01						

Table 4. Regression equations for the ratio of different soil element and zirconium

The ratios between the mobile elements Ca, P, Na and "stable" Zr clearly indicated their removal in upper layers and enrichment downward, with a minimum in the Ah horizon (Figure 6). The steeper slope indicates a faster migration of Ca compared with either P or Na. Figure 6. The ratio of Ca/Zr, Na/Zr, and P/Zr as a function of soil age and regression curves. However, literature reported that correlations between Ca, P, Na removed with time are described by linear and logarithmic models for non-calcareous soils (Lichter, 1998; Egli & Fitze, 2000; Sauer et al., 2007).

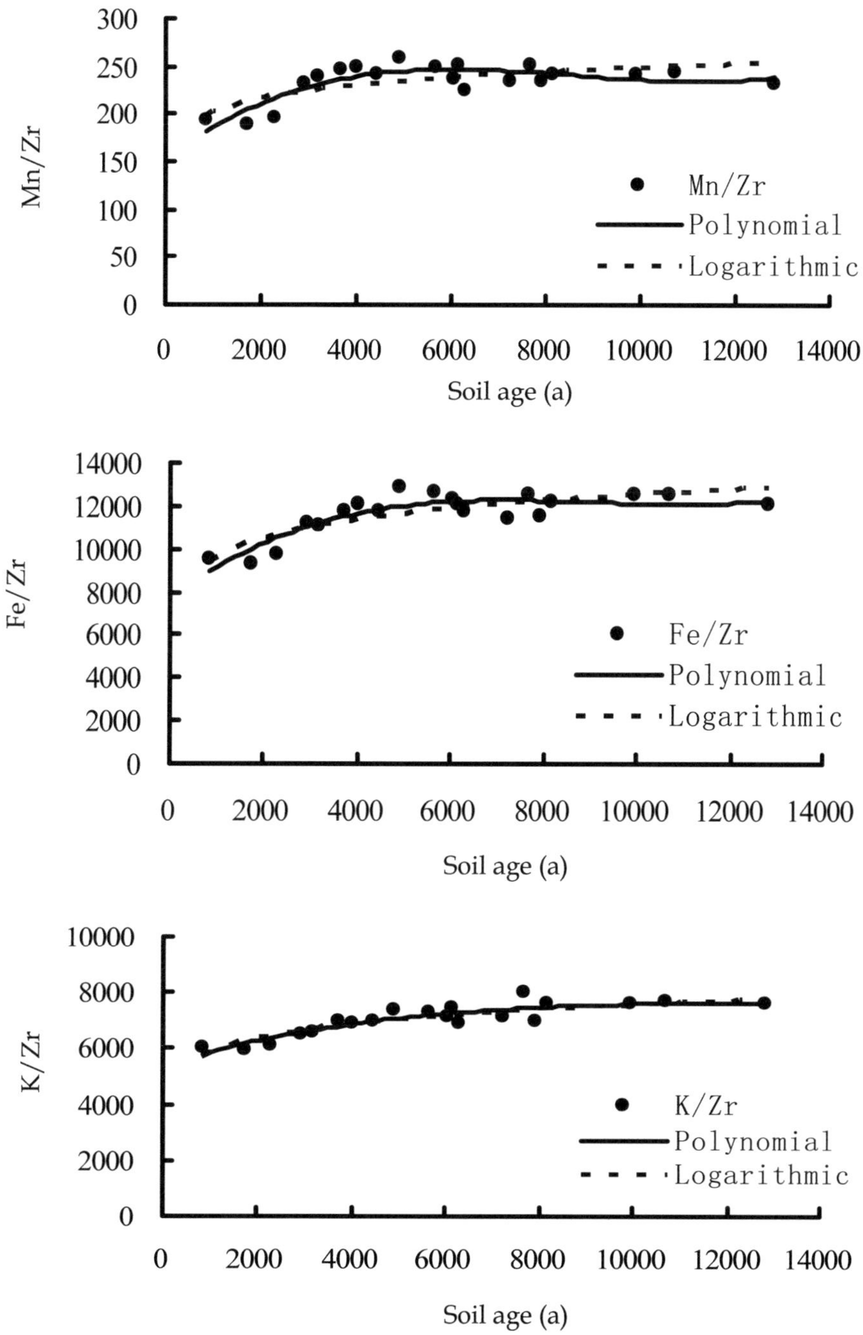

Fig. 5. The ratio of Mn/Zr, Fe/Zr, and K/Zr as a function of soil age and regression curves

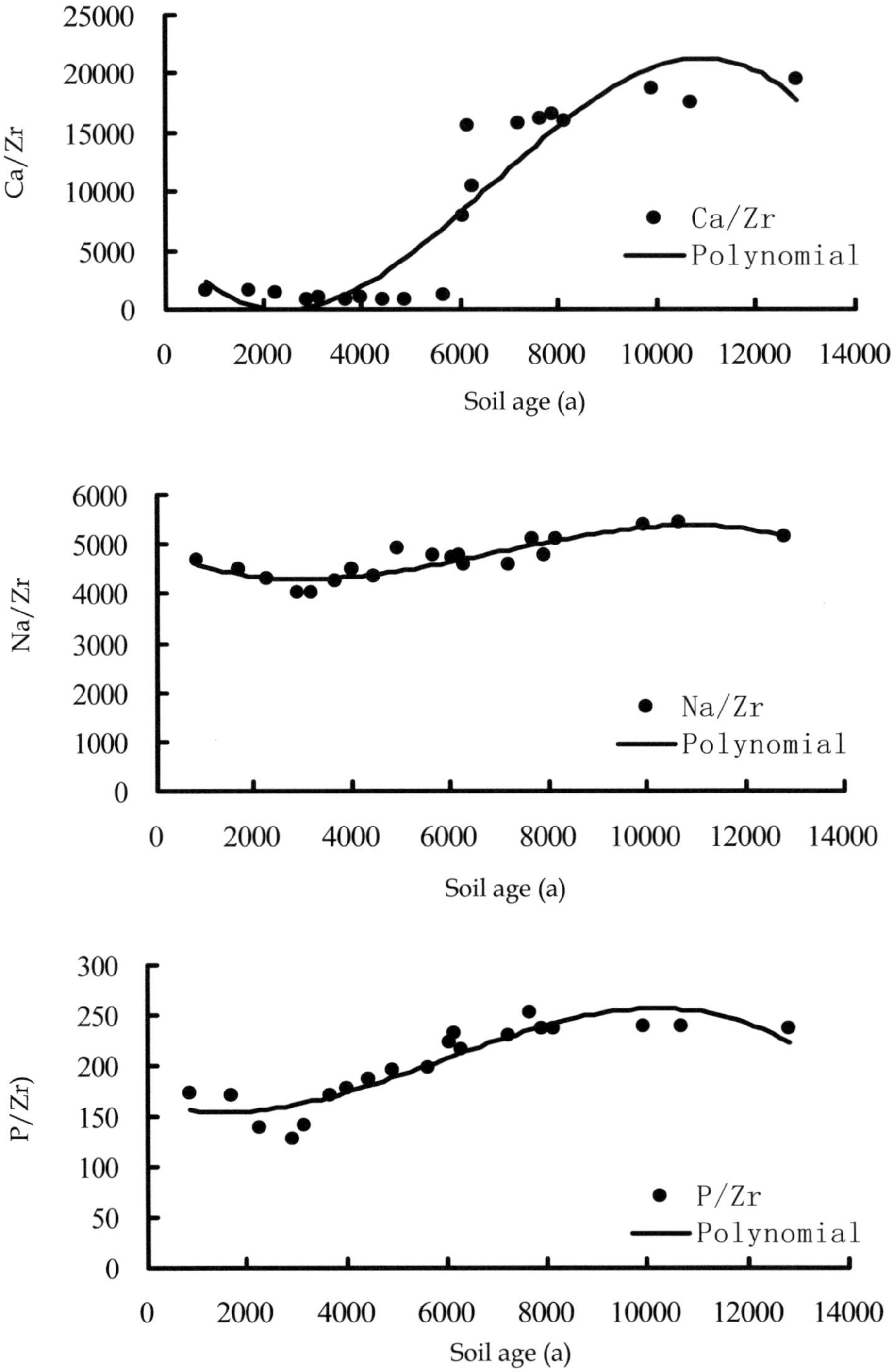

Fig. 6. The ratio of Ca/Zr, Na/Zr, and P/Zr as a function of soil age and regression curves

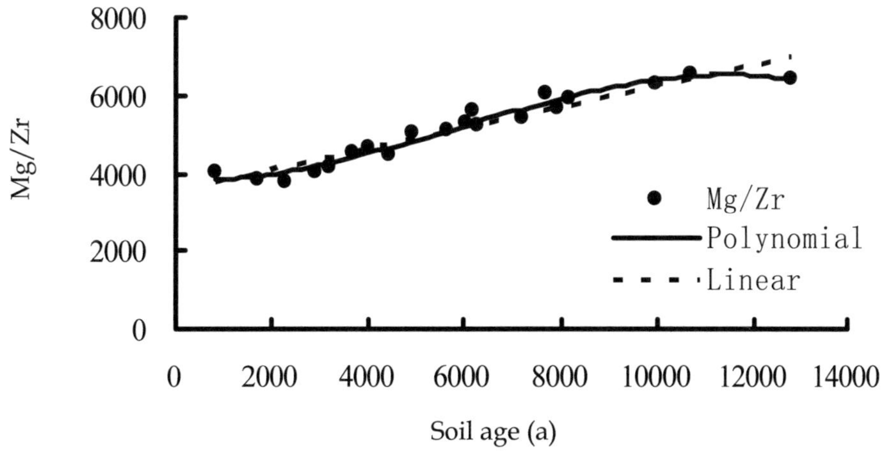

Fig. 7. The ratio of Mg/Zr as a function of soil age and regression curves

Linear and third order polynomial functions for Mg/Zr were statistically very significant, and had high r^2 (0.92 and 0.96) (Table 5). However, they showed different trends (Figure 7). The linear function indicated a constant increasing, implying a constant depletion during the Holocene, a feature difficult to accept given that the Heilu soil formed under varying climatic condition (Tang & He, 2004). In contrast, the third order polynomial function showed a rapid removal in the surface soil, followed by a slower depletion in older soils, remaining constant in the parent material. But linear and logarithmic increases for different soil chronosequences have been described in many studies elsewhere (Lichter, 1998; Egli & Fitze, 2000).

4. Conclusions

We studied soil chronofunctions, based on soil properties and ^{14}C ages in Holocene of a typical soil (Heilu) profile developed from loess in Luochuan. Linear, logarithmic, and third order polynomial functions were used to fit the relationships between soil properties and soil ages. The results indicated that the third order polynomial function was the best choice to fit the relationships between clay, silt, sand and soil ages. The trend line confirmed the presence of a humic A horizon in the profile, with higher clay and silt contents attributed to local clay formation in the relative warm and humid period between 1500 and 4200 a B.P. (Tang & He, 2004). The logarithmic functions explained best the variations of soil organic carbon and pH with time. The pH values increased with time and depth, with lower increases in older soils. The variation of $CaCO_3$ content, and ratios between Mn/Zr, Fe/Zr, K/Zr, Mg/Zr, Ca/Zr, P/Zr, and Na/Zr with soil age were better described by the use of three order polynomial functions. The results indicated that mobility of Mn, Fe and K was greater in the upper layers with depletion and enrichment of them was very weak in the lower soils. The trend of chronofunction for Mg showed that a quickly leaching in surface soil, then a slower depletion in older soils, and a constant in parent material. The ratios between Ca, P, Na and Zr clearly indicated leaching of them in upper centimeters and enrichment in lower soils, also the minimum in Ah horizon.

5. Acknowledgment

This research was jointly supported by the China Postdoctoral Science Foundation (Grant No. 20110491162), Foundation of Hubei Provincial Department of Education (Grant No. Q20111207, XD20100595), Open Research Fund Program of the State Key Laboratory of Soil Erosion and Dryland Farming on the Loess Plateau (Grant No. 10501-266), and Foundation of China Three Gorges University (Grant No. KJ2009B033, KJ2009A002).

6. References

Bain, D. C. ; Mellor, A. ; Robertson-Rintoul, M. S. E. & Buckland, S. T. (1993). Variations in weathering processes and rates with time in a chronosequence of soils from Glen Feshie, Scotland. *Geoderma*, Vol. 57, No. 3, pp. 275-293, ISSN 0016-7061

Barrett, L. R. (2001). A strand plain soil development sequence in Northern Michigan, USA. *Catena*, Vol. 44, No. 3, pp. 163-186, ISSN 0341-8162

Birkeland, P. W. (1984). Holocene soil chronofunctions, Southern Alps, New Zealand. *Geoderma*, Vol. 34, No. 2, pp. 115-134, ISSN 0016-0761

Bockheim, J. G. (1980) Solution and use of chronofunctions in studying soil development. *Geoderma*, vol. 24, No.1, pp. 71-84, ISSN 0016-0761

Bockheim, J. G. (1990). Soil development rates in the Transantarctic Mountains. *Geoderma*, Vol. 47, No. 1-2, pp. 59 -77, ISSN 0016-0761

Bronger, A. & Heinkele, TH. (1989). Micromorphology and genesis of paleosols in the Luochuan loess section, China : Pedostratigraphic and environmental implications. *Geoderma*, Vol. 45, No. 2, pp. 123-143, ISSN 0016-0761

Bronk Ramsey, C. (2001). Development of the radiocarbon calibration program. *Radiocarbon*, Vol. 43, No. 2A, pp. 355-363, ISSN 0033-8222

Carreira, J. A.; Niell, F. X. & Lajtha, K. (1994). Soil nitrogen availability and nitrification in Mediterranean shrublands of varying fire history and successional stage. *Biogeochemistry* Vol. 26, No. 3, pp. 189–209, ISSN 0168-2563

Chen, Q. Q.; Shen, C. D.; Sun, Y. M.; Peng, S. L.; Yi, W. X.; Li, Z. A. & Jiang, M. T. (2005). Mechanism of distribution of soil organic matter with depth due to evolution of soil profiles at the Dinghushan Biosphere Reserve. *Acta Pedologica Sinica*, Vol. 42, No. 1, pp. 1-8, ISSN 0564-3929

Chen, X. Y.; Wu, L. G.; Li, S. L. & Luo, L. P. (1998). Genesis and evolution of Heilu soils in the south of Wumeng in the Holocene. *Chinese Journal of Soil Science*, Vol. 29, No. 6, pp. 241-244, ISSN 0564-3945

Egli, M. & Fitze, P. (2000). Formulation of pedologic mass balance based on immobile elements: a revision. *Soil Science*, Vol. 165, No. 5, pp. 437-443, ISSN 0038-075X

Finke, P. A. & Hutson, J. L. (2008). Modelling soil genesis in calcareous loess. *Geoderma*, Vol. 145, No. 3-4, pp. 462-479, ISSN 0016-7061

Gallet, S.; Jahn, B. M. & Torii, M. (1996). Geochemical characterization of the Luochuan loess-paleosol sequence, China, and paleoclimatic implications. *Chemical Geology*, Vol. 133, No. 1-4, pp. 67-88, ISSN 0009-2541

Hedley, M. J.; Stewart, J. W. B. & Chauhan, B. S. (1982). Changes in inorganic and organic soil phosphorus fractions induced by cultivation practices and by laboratory incubations. *Soil Science Society of America Journal*, Vol. 46, No.5, pp. 970-976, ISSN 0361-5995

Hu, S. X. (1994). Genesis and evolution of heilu soils in the middle and east of Gansu Province. *Acta Pedologica Sinica*, Vol. 31, No. 3, pp. 295-304, ISSN 0564-3929

Huang, C. M. & Gong, Z. T. (2000). Quantitative studies on development of tropical soils: a case study in northern Hainan Island. *Scientia Geographica Sinica*, Vol. 20, No. 4, pp. 337-342, ISSN 1000-0690

Jacobson, G. L. & Birks, H. J. B. (1980). Soil development on recent end moraines of the Klutlan Glacier, Yukon Territory, Canada. *Quaternary Research*, Vol. 14, No. 1, pp. 87- 100, ISSN 0033-5894

Jahn, B. M.; Gallet, S. & Han, J. (2001). Geochemistry of the Xining, Xifeng and Jixian sections, Loess Plateau of China: eolian dust provenance and paleosol evolution during the last 140 ka. *Chemical Geology*, Vol. 178, No.1, pp. 71-94, ISSN 0009-2541

Jia, Y. F.; Pang, J. L. & Huang, C. C. (2004). pH value's measurement and research of its palaeoclimatic meaning in the Holocene loess section. *Journal of Shaanxi Normal University (Natural Science Edition)*, Vol. 32, No. 2, pp. 102-105, ISSN 1672-4291

Langley-Turnbaugh, S. J. & Bockheim, J. G. (1998). Mass balance of soil evolution on late Quaternary marine terraces in coastal Oregon. *Geoderma*, Vol. 84, No. 4, pp. 265-288, ISSN 0016-7061

Lichter, J. (1998). Rates of weathering and chemical depletion in soils across a chronosequence of Lake Michigan sand dunes. *Geoderma*, Vol. 85, No. 4, pp. 255-282, ISSN 0016-7061

Liu, D. S. (2009). *Loess and Arid Environment*. Anhui Science and Technology Press, ISBN 978-753-3743-10-9, Hefei, China

Maher, B. A.; Thompson, R. & Zhou, L. P. (1994). Spatial and temporal reconstructions of changes in the Asian palaeomonsoon: A new mineral magnetic approach. *Earth and Planetary Science Letters*, Vol. 125, No. 1-4, pp. 461-471, ISSN 0012-821X

Manner, H. I. & Morrison, R. J. (1991). A temporal sequence (chronosequence) of soil carbon and nitrogen development after phosphate mining on Nauru Island. *Pacific Science*, Vol. 45, No. 4, pp. 400–404, ISSN 0030-8870

Matthews, J. A. (1992). *The Ecology of Recently-Deglaciated Terrain: A Geoecological Approach to Glacier Forelands and Primary Succession*. Cambridge University Press, ISBN 978-052-1361-09-5, Cambridge, UK.

Mellor, A. (1985). Soil chronosequences on Neoglacial moraine ridges, Jostedalsbreen and Jotunheimen, southern Norway: a quantitative pedogenic approach, In: *Geomorphology and Soils*, K. S. Richards, R. R. Arnett & S. Ellis (Eds.), pp. 289-308. HarperCollins Publishers Ltd., ISBN 978-004-5510-93-1, London, UK

Mellor, A. (1987). A pedogenic investigation of some soil chronosequences on neoglacial moraine ridges, southern Norway: examination of soil chemical data using principal components analysis. *Catena*, Vol. 14, No. 5, pp. 369–381, ISSN 0341-8162

Merritts, D. J.; Chadwick, O. A. & Hendricks, D. M. (1991). Rates and processes of soil evolution on uplifted marine terraces, northern California. *Geoderma*, Vol. 51, No. 1-4, pp. 241-275, ISSN 0016-7061

Miao, X. D.; Mason, J.; Goblet, R. J. & Hanson, P. R. (2005). Loess record of dry climate and aeolian activity in the early-to mid-Holocene, central Great Plains, North America. *The Holocene*, Vol. 15, No. 3, pp. 339-346, ISSN 0959-6836

Mook, W.G.& Streurman, H.J. (1983). Physical and chemical aspects of radiocarbon dating. In: ^{14}C *and Archaeology. Proceedings of the First International Symposium (=PACT 8)*, W. G. Mook, & H. T. Waterbolk, (Eds.), pp. 31 - 55. Strasbourg.

Morozova, T. D. (1990). Relict features of paleosols formed on loess and their age. *Quaternary International*, Vol. 7-8, pp. 29-35, ISSN 1040-6182

Muhs, D. R.; McGreehin, J. P.; Beann, J. & Fisher, E. (2004). Holocene loess deposition and soil formation as competing processes, Matanuska Valley, southern Alaska. *Quaternary Research*, Vol. 61, No. 3, pp. 265-276, ISSN 0033-5894

Olson, J. S. (1958). Rates of succession and soil changes on Southern Lake, Michigan, sand dunes. *Botanical Gazette*, Vol. 119, No. 3, pp. 125–170, ISSN 0006-8071

Reimer, P. J.; Baillie, M. G. L.; Bard, E.; Bayliss, A.; Beck, J. W.; Bertrand, C. J. H.; Blackwell, P. G.; Buck, C. E.; Burr, G. S.; Cutler, K. B.; Damon, P. E.; Edwards, R. L.; Fairbanks, R. G.; Friedrich, M.; Guilderson, T. P.; Hogg, A. G.; Hughen, K. A.; Kromer, B.; McCormac, G.; Manning, S.; Bronk Ramsey, C.; Reimer, R. W.; Remmele, S.; Southon, J. R.; Stuiver, M.; Talamo, S.; Taylor, F. W.; Van der Plicht, J. & Weyhenmeyer, C. E. (2004). IntCal04 terrestrial radiocarbon age calibration, 0–26 cal kyr BP. *Radiocarbon*, Vol. 46, No. 3, pp. 1029–1058, ISSN 0033-8222

Piccolo, M. C.; Neill, C. & Cerri, C. C. (1994). Natural abundance of ^{15}N in soils along forest-to-pasture chronosequences in the western Brazilian Amazon Basin. *Oecologia*, Vol. 99, No. 1-2, pp. 112–117, ISSN 0029-8549

Sauer, D.; Schellmann, G. & Stahr, K. (2007). A soil chronosequence in the semi-arid environment of Patagonia (Argentina). *Catena*, Vol. 71, No. 3, pp. 382-393, ISSN 0341-8162

Schaetzl, R. J.; Barrett, L. R. & Winkler, J. A. (1994). Choosing models for soil chronofunctions and fitting them to data. *European Journal of Soil Science*, Vol. 45, No. 2, pp. 219-232, ISSN 1351-0754

Tang, K. L & He, X. B. (2002). Revelation on genesis of multi paleosol from Quaternary loess profile. *Acta Pedologica Sinica*, Vol. 39, No. 5, pp. 609-617, ISSN 0564-3929

Tang, K. L. & He, X. B. (2004). Re-discussion on loess-paleosol evolution and climatic change on the Loess Plateau during the Holocene. *Quaternary Sciences*, Vol. 24, No. 2, pp. 129-139, ISSN 1001-7410

Vincent, K. R.; Bull, W. B. & Chadwick, O. A. (1994). Construction of a soil chronosequence using the thickness of pedogenic carbonate coatings. *Journal of Geological Education*, Vol. 42, 316-324, ISSN 0022-1368

Vogel, J. S.; Southon, J. R. & Nelson, D. E. (1987). Catalyst and binder effects in the use of filamentous graphite for AMS. *Nuclear Instruments and Methods in Physics Research Section B: Beam Interactions with Materials and Atoms*, Vol. 29, No. 1-2, pp. 50-56. ISSN 0168-583X

Zarin, D. J. & Johnson, A. H. (1995). Base saturation, nutrient cation, and organic matter increases during early pedogenesis on landslide scars in the Luquillo Experimental Forest, Puerto Rico. *Geoderma*, Vol. 65, No. 3-4, pp. 317–330, ISSN 0016-7061

Zhao, J. B.; Hao, Y. F. & Yue, Y. L. (2006). Change of paleosol and climate during middle Holocene in Luochuan area of Shaanxi Province. *Quaternary Science*, Vol. 26, No. 6, pp. 969-975, ISSN 1001-7410

Zhu, X. M.; Li, Y. S.; Peng, X. L. & Zhang, S. G. (1983). Soil of the loess region in China. *Geoderma*, Vol. 29, No. 3, pp. 237-255, ISSN 0016-7061

4

Radiocarbon Dating in Archaeological Sites Chronology

Danuta Michalska Nawrocka[1], Małgorzata Szczepaniak[1] and Andrzej Krzyszowski[2]
[1]*Adam Mickiewicz University, Institute of Geology,*
[2]*Archaeological Museum,*
Poznań,
Poland

1. Introduction

Charcoal and bones are materials commonly used for the radiocarbon dating in geological and archaeological research. The difficulties with ^{14}C dating of charcoal and wood may be associated with the origin of analysed fragments, the conditions of the sediments, such as pH or humidity. As far as bones are concerned, the difficulties with radiocarbon dating may be connected with their state of preservation, collagen presence and the possible contamination by carbon from other sources.

In some cases, despite the standard method of chemical pretreatment, it may be necessary to customise the applied treatment to the investigated material. The results of radiocarbon dating of bones and wood were compared with the relative chronology established by archaeologists, based usually on the typology of artefacts or pottery.

The presented results of analyses refer to the territory of Wielkopolska (Great Poland), including among others, prehistoric and early medieval settlement sites in Suchy Las, Łęki Wielkie, Szczodrzykowo, Trzcielin, Snowidowo, Żerniki and Zielęcin (fig. 1). Samples of bones and pieces of wood selected for dating come from different periods of time and various cultures distinguished in archaeology, from the Funnel Beaker culture and the Lusatian culture to the developed phases of the Early Middle Ages (fig. 2). The history of the settlement in Wielkopolska, a historical Polish district comprising the basin of the central Warta River, is quite complex (Kobusiewicz, 2008). The origins of occupation in the area, recognised by archaeologists, date back to the late Paleolithic, about 12-10 millennia BC. The Mesolithic and Neolithic hunter-gatherer communities lived here in the period between 9 and 3 millennium BC. The first agricultural societies appeared in the Neolithic, namely 6-3 millennium BC (the Linear Pottery culture, the Funnel Beaker culture, the Globular Amphora culture, the Corded Ware culture). They were followed by communities of the early Bronze Age (the Iwno culture, the Únětice culture and the Trzciniec culture), the developed Bronze Age societies (the Lusatian culture), early Iron Age cultures (the Pomeranian, the Jastorf and the Przeworsk), the cultures of the period of the Roman influences (the Przeworsk and the Wielbark), and finally the early Medieval settlement, when the Polish statehood began to emerge.

Some of these archaeological cultures, such as Lusatian or Przeworsk, are characteristic only for the selected area of Central Europe, among others the Czech Republic, Poland and Germany.

Poznań, which currently is the capital of Wielkopolska, obtained city rights in 1253, but it fulfilled the role of a centre, the most import stronghold already since the times of Duke Mieszko I of the Piasts dynasty, in the end of the 10th century.

For the majority of the presented sites from the Wielkopolska Province, ^{14}C dates have been collected for the first time (with the exception of site 4 in Łęki Wielkie) and they have been intended to help determine the chronology, possible phases of development or point to the beginnings of given strongholds.

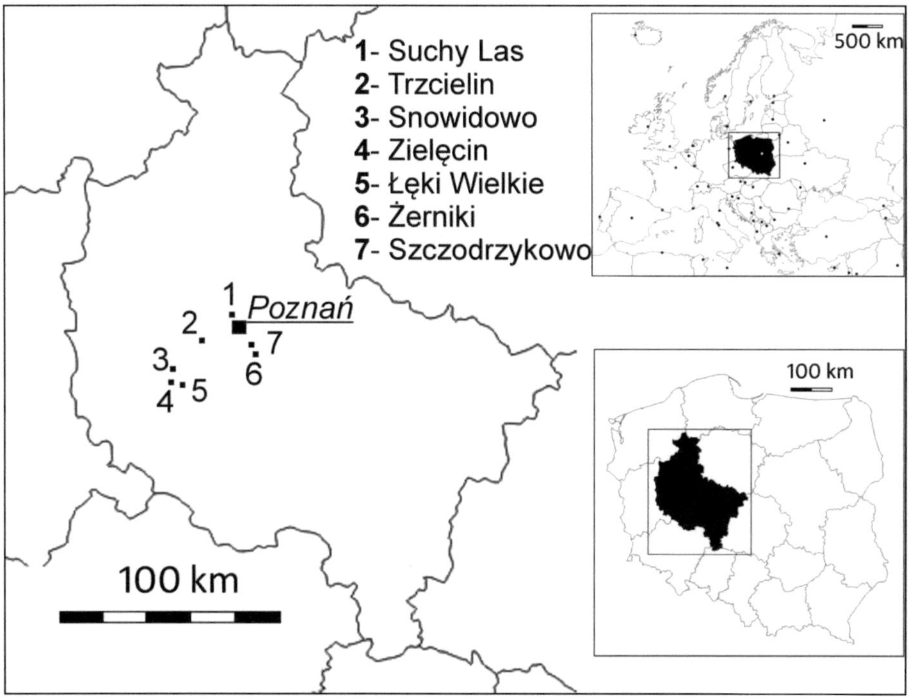

Fig. 1. The studied area against the background of Europe, Poland and the Wielkopolska Province.

2. Archaeological background

The dated samples come from a few archaeological sites from the Wielkopolska Province, situated in the vicinity of Poznań. They are located in the area of Trzcielin, Suchy Las, Żerniki, Zielęcin, Snowidowo, Łęki Wielkie and Szczodrzykowo (fig.2). These sites are chronologically diversified and some of them yielded a number of chronologically and culturally varied developmental phases. However, the presence of the Early or Late Middle Ages on the majority of them, is a common trait. Remains of the Przeworsk culture are less frequent here. Some of the cultures present in the analysed area are characteristic only for Central Europe.

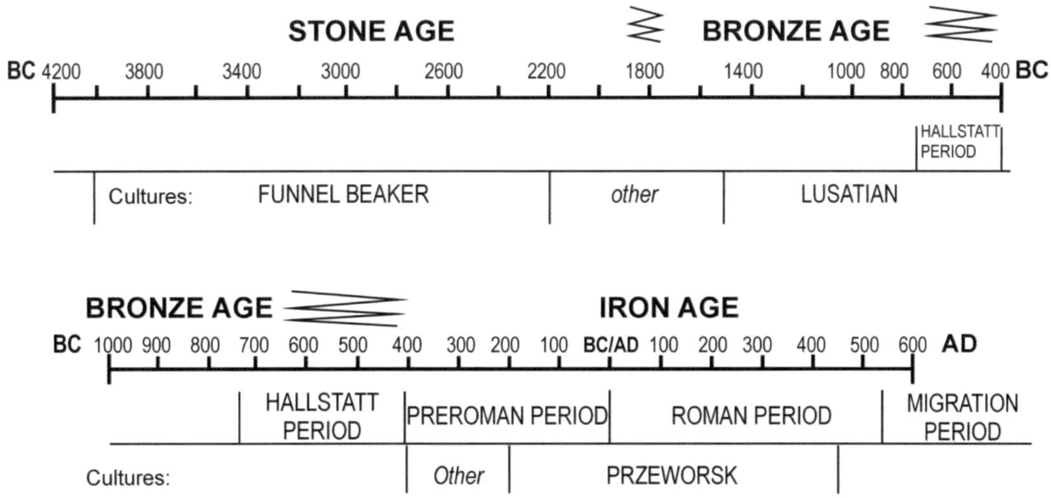

Fig. 2. Chronology and archaeological cultures of the Wielkopolska region discussed in the paper (*other cultures* - cultures occurring in the analysed region, from which there were not any samples available).

2.1 Suchy Las

Archaeological excavations conducted in 2007 at site 6 in Suchy Las near Poznań produced archaeological material from five chronological horizons (basing on relative chronology) related to:

- the communities of the Funnel Beaker culture from the Neolithic (from about 3200 until about 2500 BC),
- the communities of the Pomeranian culture from the Early Iron Age (from about 400 BC till the turn of the eras),
- the Early Middle Ages - from the 13th - 14th century,
- the Late Middle Ages - from the end of the 14th - the beginning of the 16th century
- the modern period - from the 17th -19th century.

The early medieval village of *Zuchilecz* (Suchy Las) was first mentioned in written sources in 1170 (Długosz, 1874). It was suggested that the village had been granted by Mieszko the Old and given to the order of St. John of Jerusalem, who had run a hospital of St. John of Jerusalem in Poznań in the 2nd half of the 12th century.

2.2 Trzcielin

Archaeological Site 63 in Trzcielin is located in the district of Poznań. The site yielded an extensive early medieval settlement, recognised mainly on the basis of artefacts. Apart from the Early Middle Ages, that can be probably divided into two phases, and the modern period, (the 17th -beginning of the 20th century), the 2009 excavations produced also a prehistoric horizon, probably dating to the Neolithic (second half of the 2nd millennium BC). For dating purposes, a sample from feature 64 was collected. Archaeologists date this feature back to the Middle Ages.

For these kind of features the obtained date is significant in terms of the possibility of chronological verification of the development of feudal relations in Wielkopolska. In this case, it was a subordinated production settlement site, where significant quantities of turf ore were smelted, also for its inhabitants. It was also an important settlement site in view of the needs of a nearby central stronghold of Poznań, as a duke's superior centre.

2.3 Snowidowo

Excavations at site number 9 in Snowidowo, carried out in 2009, produced four settlement complexes:

- a trace of prehistoric occupation (probably from the Neolithic),
- a settlement site of the Przeworsk culture from the late pre-Roman period,
- a trace of occupation from the Early Middle Ages (from phase F),
- a settlement point from the modern period.

There is also the hypothesis of possible distinguishing between different development phases of the Przeworsk culture, a pre-Roman (from 2nd cent. BC till BC/AD) and a Late Roman settlement.

2.4 Zielęcin

A previously unknown archaeological site 15 in Zielęcin was discovered in the course of excavations conducted in 2009. It yielded modern features and features dating back to the early phases of the Early Middle Ages. Based on the production techniques of pottery, part of the settlement site from which the samples were collected was dated to the 6th and 7th century A.D. Therefore it is an assemblage related exclusively to A-A/B type of the early medieval pottery in the Pomerania region according to the classification of Łosiński & Rogosz (1983, 1986). Almost all pottery is hand-formed (without the usage of a potter's wheels) with no decoration. In view of its early chronology, within the turn of Antiquity and the Early Middle Ages, the scientific value of the discovered site is enormous. It is also related to the question of the appearance of the earliest assemblages with the Slavic pottery in the area of Wielkopolska.

2.5 Łęki Wielkie

Site number 4 in Łęki Wielkie was discovered in 1933 and verified in 1987. The most recent excavations were conducted in 2009. The following cultural units have been registered:

- a settlement site of the Przeworsk culture,
- a settlement site accompanying a stronghold from the Early Middle Ages (from phase C),
- a trace of occupation from the Late Middle Ages (15th -16th century),
- a trace of occupation from the modern era (17th - beginning of the 20th century).

The sample selected for dating purposes came from a settlement accompanying a stronghold from the end of the 9th century till about the 1st half of the 10th century AD (Early Middle Ages, phase C), basing on relative chronology.

Radiocarbon dating was supposed to determine the chronological relationship between the settlement site and the nearby early medieval stronghold, representing a moment of the transition between the tribal period and the emerging early Piast state, interesting for the history of Wielkopolska. An extensive settlement site near the stronghold, several dozen hectares in size, has been tentatively dated (only on the basis of the traits of the pottery) to the period from the end of the 9th century to about 1st half of the 10th century (Kara & Krąpiec, 2000).

According to the sources obtained from survey excavations in the 1980s, the nearby stronghold was analogously dated. The ^{14}C date was intended to confirm unambiguously the dating determined exclusively on the basis of the typology of archaeological sources.

2.6 Żerniki

Archaeological site in Żerniki was discovered and excavated in 1986. Excavations, continued in 2009 due to a local industrial investment, produced four chronological horizons (fig. 2):

- a trace of occupation from the Neolithic (2nd half of the 2 millennium BC),
- a settlement site of the Przeworsk culture (4th century AD),
- a settlement point from the Early Middle Ages (12th -13th century AD),
- and a settlement point from the modern period (17th – beginning of the 20th century).

The most valuable materials from the site are related to the Przeworsk culture. On the site the greatest number of features were registered and the rescue excavations were aimed at recognising as much of the spatial range of the settlement site as possible. Both features chosen for dating purposes come from horizons related to the settlement of the Przeworsk culture (3rd -4th century AD). Although a great number of features dating back to this period were excavated, there were not enough sources for the precise determination of the time of its existence. The possibility of determining the chronological variability within the same settlement site was also a significant aspect of the conducted studies.

Apart from one Roman coin (follis) from the times of emperor Constantine I and two fibulae (type A.158 and A.162 (Almgren, 1923), excavated outside features, it was not possible to determine the chronology of the majority of open elements of this settlement site (such as lime kilns, postholes or heaps of fired lime).

2.7 Szczodrzykowo

The analysed site is located in the commune of Kórnik, south-east of Poznań. The finds in the area of site number 4 in Szczodrzykowo point to the existence of a multicultural settlement site here, namely (fig. 2):

- a Stone Age trace of occupation,
- a settlement site of the Lusatian culture,
- a settlement site of the Przeworsk culture,
- a settlement site from the early Middle Ages,
- a trace of occupation from the modern era.

3. Sample description

The material from the area of Wielkopolska collected for dating purposes includes mostly samples of wood and fragments of human and animal bones excavated at sites, in archaeological layers of varied chronology (Fig.2, 3 and 4).

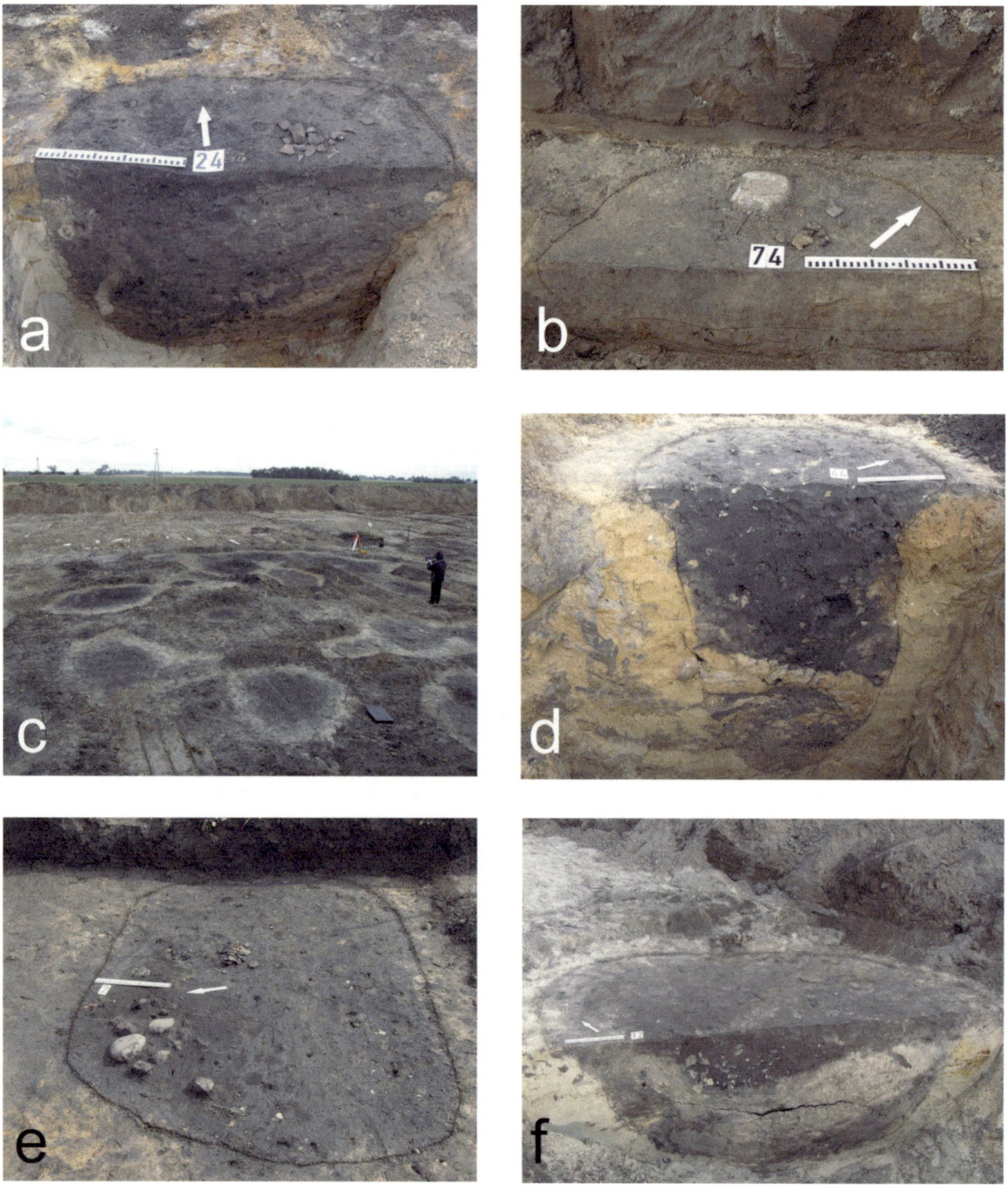

Fig. 3. Chosen archaeological sites: a) Suchy Las, feature 24 – storage pit that contained one of the samples of bones for dating; b) Snowidowo, feature 74 that contained the dated bone; c) Trzcielin – general view of excavations; d) Trzcielin - feature 64, from which a sample of an animal bone was collected for dating; e) Zielęcin, feature 7 – half dugout from which a sample for ^{14}C measurement was collected; f) Łęki Wielkie, feature 12 – a dwelling structure – half dugout, from which a bone was collected for dating.

Fig. 4. Chosen archaeological sites; a) Żerniki - feature 218 – a dwelling structure from which the first bone sample was collected; b) Żerniki - feature 302 – a storage pit, from which the second bone sample was collected c) Szczodrzykowo, wells of varied age with wooden lining, from which wood samples were collected for dating - feature 52 (late medieval); d) Szczodrzykowo – feature 56 (the Lusatian culture).

3.1 Suchy Las

Two samples from the chronological horizon related to the Middle Ages were selected for radiocarbon dating:

- a sample dated by archaeologists on the basis of relative chronology to the period from the end of the 10th century till the 1st half of the 11th century, (phase D2);
- the period from the 2nd half of 11th century till 1st half of the 12th century (phase E2).

Both samples are animal bones from features 24 and 26. Feature 24 is a storage pit, whereas feature 26 is a dwelling structure, a half dugout (fig.3a).

3.2 Trzcielin

A sample selected for ^{14}C dating was procured from a part of the site estimated by archaeologists to the Early Middle Ages (late 10th century to the turn of the 11th and 12th century). It was an animal bone chosen out of more than 1,200 bones from feature 64. The function of this feature has not been fully recognised. In the first phase it was probably a bloomery, whereas once the activities relating to the smelting (or processing) the iron raw material were abandoned, it began to fulfil the role of a rubbish pit.

3.3 Snowidowo

In terms of the chronology, a sample chosen for dating came from part of the site related to the Przeworsk culture, preliminarily estimated by archaeologists to the pre-Roman Period. The assemblage of finds from this chronological period numbered, among others, 84 features and only 157 potsherds. A chosen sample of an animal bone was collected from feature 74.

3.4 Zielęcin

Samples selected for ^{14}C measurements are animal bones from features 6 and 7. In both cases they were excavated within the dwelling structures (half dugouts), in the area of an early medieval settlement site. A phase of this part of the settlement was preliminary dated to the period of the Middle Ages (phase A-B) on the basis of the production techniques of the pottery. In terms of archaeological sources it is a period between the end of the 6th and 7th century AD.

3.5 Łęki Wielkie

The material chosen for dating from this site is an animal bone from a dwelling structure - a half dugout. The site is a settlement accompanying the stronghold (phase C of the Middle Ages) estimated by archaeologist on the basis of relative chronology to the period from the end of the 9th century till about the 1st half of the 10th century AD.

3.6 Żerniki

Both features selected for dating are dated to the horizons related to a settlement site of the Przeworsk culture (3rd - 4th century AD). For ^{14}C analyses two animal bone samples were chosen: from feature 218 - the only dwelling structure, and feature 302 - a storage pit that contained a glazed glass bead imported from the territory of the Roman provinces.

3.7 Szczodrzykowo

The material selected for radiocarbon dating comes from two features. The first one (fig. 4d), feature 56, represents a settlement site of the Lusatian culture (from the Hallstatt period, namely from the 7th -5th century BC). It is a fragment of wood from a well with a wooden lining made of a burnt tree trunk. The other sample is a wooden fragment from feature 52, qualified to a settlement site from the Late Middle Ages (14th century, fig. 4c)). The sample has also been taken from a wood-lined well, however in this case the well was erected by stacking logs one on top of another and overlapping them at the corners.The age ranges for both features were established on the basis of archaeological research and absolute dating verification is needed.

4. Geological background

The studied archaeological sites are located in the central-western part of Poland (Wielkopolska region). This area is characterized by postglacial relief and its morphological features were shaped by glacial and fluvial processes during the Weischelian glaciation in late Pleistocene and Holocene. Consequently, late Quaternary deposits in the Wielkopolska, specifically in the vicinity of Poznan, are dominated by glacial tills, fluvioglacial sands and gravels as well as alluvia of the Warta river. The former are exposed in the end morains of the Poznan phase and the latter two are encountered in the Warta river marginal valley (Stankowski, 1996).

^{14}C dating was performed on the fragments of bones, excavated from mineral substratum, mentioned above. In the Suchy Las and Trzcielin sites the material for age determination was found within loamy sands, however, in the latter the sand co-occurred with glacial till. In the Łęki Wielkie, Żerniki and Snowidowo, the parent material consisted of fine-grained sand. In Snowidowo the sand laterally changed to glacial till. In Zielęcin the bone fragments were extracted from sandy till and in Szczodrzykowo - from sandy till surrounded by fine sands.

The sampled material was evidently influenced by physical and chemical processes operating in agricultural landscape. The area where study sites are located, has been extensively used for agricultural purposes and fertilization of the soils. Migration of soil solutions led to secondary changes within the fossil bones, and consequently, to inaccurate radiocarbon age.

5. Methods

Wood and bone samples, chosen from each site, were radiocarbon dated using the technique of AMS in The Poznań Radiocarbon Laboratory. The laboratory is equipped with a accelerator mass spectrometer type 1,5 SDH-Pelletron, Model „Compact Carbon AMS".

Fragments of charcoal and wood are most frequently used materials for radiocarbon dating. However, the results of dating wood and charcoal do not always reflect the time of cutting down a tree. A tree grows successively, building on subsequent rings each year. Depending on whether the analysed fragment comes from a central part of a trunk or a young twig, or whether it is a group of mixed charcoal fragments collected from an archaeological layer (McFadgen, 1982; Ashmore, 1999), there might occur significant discrepancies between the radiocarbon age and the age of the relevant archaeological level.

In the context of archaeological sites, the likely time lapse between the moment of cutting down a tree and using it for erecting structures, or the possibility of re-using building materials seems to have much less influence.

The preparation protocol for individual samples is important because of different preservation stage, sample size and possible samples loss during treatment (Nawrocka et al., 2007, 2009; Rebollo et al., 2008; Szczepaniak et al., 2008). The first stage of wood and charcoal purification was removal of macroscopically visible plant roots. Then the samples have been subjected to the "acid-base-acid" (ABA) preparation step by treatment with 1M HCl solution. The aim of this step was to dissolve carbonates and other soluble minerals. The second stage was submerging the samples in the base solution (0,025 M NaOH and 0,5 M NaOH) to remove humic compounds. The last ABA step was the repeated acid treatment (0,25 M HCl). Each time, the samples were bathed in deionised water to restore the neutral pH.

Subsequently the wood material was subjected to a bleach pretreatment with sodium chlorite ($NaClO_2$) to remove lignins, resins, waxes. In case of very small samples or poorly preserved fragments the temperature and the time of each step was individually match. In the next stages of samples preparation to dating the combustion in vacuum to obtain carbon dioxide from organic material and reduction were used. Finally, graphite with iron was pressed to form a cathode ready for AMS measurement.

The age of bones excavated at archaeological sites is usually closely related with the period of its functioning. The content of material relevant for dating, e.g. collagen in case of bones, depends on their age, state of preservation or depositional environments. Remarks on the state of collagen preservation are given in Table 1. One of the major mistakes which occurs while dating bones is the contamination by humus acids and carbonates of a different origin. At the same time the pretreatment, aimed at eliminating contamination, may cause partial loss of material, as the collagen is susceptible to the effect of concentrated solutions. Almost all samples contained enough collagen for dating.

If the amount of collagen extracted from the sample is lower than 1% of the starting weight of the bone material used, the sample is usually rejected prior to dating.

During the process of transformation of the sample to CO_2 in the laboratory, the reliability of dating the sample is controlled by various parameters. The carbon content of the product should be between 30 and 50% of the weight of the collagen. Values beyond this range are indicative of contamination or degradation. The C:N atomic ratio should be between 2.9 and 3.5 (van Klinken, 1999). Samples with higher ratios may have been contaminated with exogenous carbon and bones with lower ratios were prone to degradation. Both, inadequate carbon contents and C/N ratios result in rejection of the analyzed samples from dating.

After analysis of the percent carbon content and C:N atomic ratio in bones, in the Poznań Radiocarbon Laboratory routine bone pretreatment procedure involves (Brock et al., 2010; Brown at al., 1988, Stanley, 1990):

- ABA pre-treatment,
- gelatinization of the sample,
- ultrafiltration of the collagen.

6. Results and discussion

^{14}C dating results for the bones and wood fragments samples are given in Table 1 and 2, and calibrated ^{14}C dates are shown in figure 5. A detailed interpretation of the research material is given below, in sub-chapters relating to particular archaeological sites. The archaeological context presented in table 1, was established on the basis of relative chronology made during the excavation. Till 2011 almost in all of the sites presented in this paper relative dating based on ceramic residue, stone objects or other artefacts has been made.

6.1 Suchy Las

^{14}C dates collected from both features (animal bones) correspond relatively well with the conventional chronology determined on the basis of technical and decorative traits of pottery excavated at the site (table 1, 2; fig. 5). A rare agricultural tool discovered in one feature (24) or an imported graphite pottery found in another one can also be dated to the early Piast period, namely the transition phase between D2/E1 phases (Łosiński & Rogosz, 1983, 1986) of the Early Middle Ages. Owing to the obtained radiocarbon dates, the dates known from written sources regarding the beginnings of the early medieval village of *Zuchilecz* (Suchy Las - the first one is supposed to have existed already in 1170), has partially been confirmed. The ^{14}C dates from archaeological materials allow for the conclusion that an early medieval settlement site had existed in this place a bit earlier than historical dates suggest.

Site	Sample name	Material/ comments	Lab code	Archaeological context
Suchy Las	SLAS/ob24/35/2007	Animal bone/ 2.3%N 7.2%C	Poz-43114	a settlement site, phase D2 Early Middle Ages: 2nd half of the 10th century – 1st half of the 11th century;
	SLAS/ob26/37/2009	Animal bone/ 3.2%N 8.8%C	Poz-43115	a settlement site, phase E2 Early Middle Ages: end of 12th - 1st half of the 13th century
Trzcielin	TRZ/ob64/49/2009	Animal bone/ 2.6%N 8.4%C	Poz-43113	a settlement site, Early Middle Ages (end of the 10th-11th /12th century); a bloomery/rubbish pit
Snowidowo	SNOW/ob74/16/2009	Animal bone/ 2.4%N 7.4%C	Poz-43123	the Przeworsk culture, pre-Roman period 2nd century BC - BC/AD); rubbish pit; probably two development phases
Zielęcin	ZIEL/ob6/4/2009	Animal bone/ 1.4%N 6.6%C	Poz-43119	a settlement site, phase A-B Early Middle Ages (end of 6th -7th century); half dugout
	ZIEL/ob7/5/2009	Animal bone/ 1.4%N 5.9%C	Poz-43120	
Łęki Wielkie	LEWIEL/ob12/4/2009	Animal bone/ 2.5%N 8.3%C	Poz-43121	a settlement site accompanying the stronghold, phase C Early Middle Ages (end of 9th century to about 1st half of the 10th century); half dugout
Żerniki	ZERN/ob218/76/2009	Animal bone/ 1.2%N 4.7%C	Poz-43116	a settlement site, the Przeworsk culture (3rd – 4th century AD)
	ZERN/ob302/94/2009	Animal bone/ 1.3%N 5.7%C	Poz-43117	
Szczodrzykowo	SZCZ/ob56/40/2004	Wood - Pinus	Poz-43124	a settlement site, the Lusatian culture, Hallstatt period (7th – 5th century BC)
	SZCZ/ob52/36-37/2004	Wood - Quercus	Poz-43125	a settlement site, Late Middle Ages (14th century)

Table 1. The sample description together with the place of sample collection; Lab code: Poz—sample dated with the AMS technique in the Poznan Radiocarbon Laboratory; %N- percentage of nitrogen; %C - percentage of carbon in the sample.

Sample name	Lab code	^{14}C age BP	Calibrated age AD	
			68%	95%
SLAS/ob24/35/2007	Poz-43114	1020±30	991 (68.2%) 1026	902 (1.9%) 915 969 (88.5%) 1045 1095 (4.1%) 1120 1141 (1.0%) 1148
SLAS/ob26/37/2009	Poz-43115	895±30	1049 (27.7%) 1085 1123 (9.5%) 1138 1151 (27.0%) 1188 1199 (3.9%) 1206	1040 (39.1%) 1110 1116 (56.3%) 1215
TRZ/ob64/49/2009	Poz-43113	975±30	1020 (32.9%) 1046 1092 (28.0%) 1121 1140 (7.3%) 1148	1014 (95.4%) 1155
SNOW/ob74/16/2009	Poz-43123	1685±30	266 (4.0%) 272 335 (64.2%) 408	257 (16.9%) 299 318 (78.5%) 422
ZIEL/ob6/4/2009	Poz-43119	1510±30	539 (68.2%) 600	435 (15.0%) 491 509 (1.2%) 518 529 (79.2%) 623
ZIEL/ob7/5/2009	Poz-43120	1235±30	694 (30.0%) 748 765 (29.8%) 820 842 (8.4%) 860	687 (95.4%) 879
LEWIEL/ob12/4/2009	Poz-43121	1100±30	897 (25.6%) 923 941 (42.6%) 985	887 (95.4%) 1014
ZERN/ob218/76/2009	Poz-43116	1750±30	242 (17.8%) 265 272 (50.4%) 335	216 (95.4%) 390
ZERN/ob302/94/2009	Poz-43117	1730±30	255 (68.2%) 345	240 (95.4%) 391
SZCZ/ob56/40/2004	Poz-43124	2625±30	815 BC (68.2%) 791 BC	836 BC (95.4%) 771 BC
SZCZ/ob52/36-37/2004	Poz-43125	580±30	1317 (46.0%) 1354 1389 (22.2%) 1408	1300 (63.4%) 1369 1381 (32.0%) 1419

Table 2. Results of ^{14}C dating. Lab code: Poz—sample dated with the AMS technique in the Poznan Radiocarbon Laboratory; Calibrated age from OxCal v4.1.7 (Bronk Ramsey 2010); Atmospheric data from Reimer et al., (2009).

6.2 Trzcielin

The data obtained from feature 64 (animal bone) in the interval 1014AD (95.4%) 1155AD - can proves that the process of early feudal dependencies formation: the central stronghold of Poznań and an accompanying production settlement site, began to emerge in this region of Wielkopolska much earlier than hitherto believed. This kind of relations prevailed already in the first half of the 11th century.

6.3 Snowidowo

The obtained date from the animal bone from feature 74 verify earlier hypotheses that we can distinguish between two developmental phases of the settlement site of the Przeworsk culture (*see section* 2.3 Snowidowo), namely a pre-Roman settlement site (from the 2nd century BC-BC/AD, on the basis of relative chronology), and a Late Roman settlement site (the period between 318 AD to 422 AD, on the basis of a ^{14}C date).

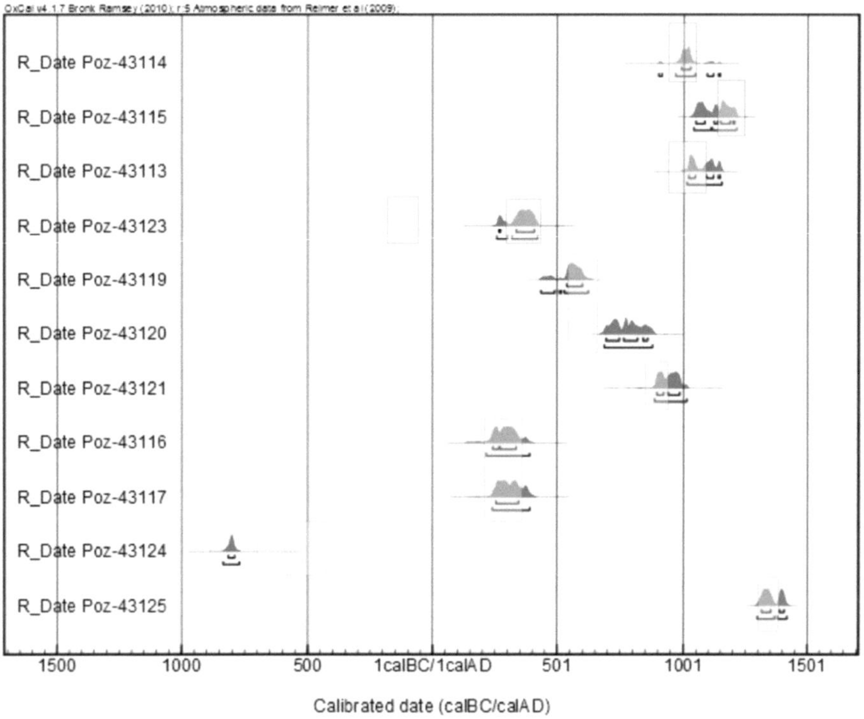

Fig. 5. Results of calibration in graphical form obtained with OxCal v4.1.7 (Bronk Ramsey, 2010; Reimer et al., 2009); the intervals of archaeological estimations are marked in grey color.

6.4 Zielęcin

^{14}C date obtained from feature 6 confirms the preliminary age estimations of the complex, whereas the other date, from an adjoining feature 7 is 'moved' at least one century in relation to the above mentioned chronology. It possibly results from the existence of two early medieval chronological phases of the settlement. The first one could have existed between the 6th -7th century, and the other functioned between the 7th -9th century. A standard (conventional) analysis of archaeological material did not capture the latter date. We also have to take into consideration that despite applied pretreatment the sample subject to radiocarbon dating could have been contaminated by carbon of a different origin.

6.5 Łęki Wielkie

The obtained ^{14}C date from feature 12 at the settlement (from 887 AD to 1014 AD) proved that, as it was expected on the basis of archaeological investigations, both settlement structures, namely a settlement accompanying the stronghold and the stronghold, functioned at the same time. Moreover, it seems that both settlement complexes were incorporated into the structures of the early Piast state in a peaceful way, which is also indirectly proven by the fact that survey excavations did not register any burnt layers either in the stronghold or in the features on the settlement site.

6.6 Żerniki

Two similar ^{14}C dates from features registered on distant outskirts of the settlement prove that the excavated structures of the settlement site come from the same chronological period, namely from about 216 AD to about 390 AD. The result of radiocarbon dating was consistent with research expectation formulated in the phase of fieldwork. Owing to ^{14}C dating, the final study on the material remains from this settlement will be devoid of speculations on the chronology of the majority of features that did not contain any artefacts.

6.7 Szczodrzykowo

The date obtained from wood from feature 52 is consistent with conventional dating determined in the course of analysis of potsherds from this feature. In case of dating the wood from feature 56 it may be concluded that preliminary established by archaeologists age ranges dating of this settlement complex should be corrected on the basis of radiocarbon dating. The reason is that the dating based on the characteristics of pottery (73 potsherds were registered) situated this feature in the Hallstatt period (between the 7th and 5th century BC). In this case, on the basis of radiocarbon dating, we should move the feature 'up' the chronological table of the Lusatian culture and date it to the times at least from 836 BC to 771 BC. Reasuming - for reliable dating of features such as the mentioned well of the Lusatian culture, it is necessary to acquire diverse material for dating and to compare the relative chronology with the results of isotope or other dating methods..

7. Conclusion

The radiocarbon method has a very wide application in various fields of science. On the basis of the presented results of the dating of bones and wood, we presented the use of this method in determining the chronology of archaeological sites. At the same time, the possibility of using various materials, in this case bones and wood, with a small quantity required for the dating of the fraction, makes it a valuable source of information for a comparison with the relative chronology.

The results of radiocarbon dating can help us verify the time intervals of given settlements or features established by archaeologists on the basis of the typology and decoration of excavated ornaments or pottery.

he comparison of the relative chronology with the results of ^{14}C measurements enabled, e.g. for the site of Zielęcin, samples from feature 6 (Poz-43119), to confirm the archaeological expectations. The result of dating the sample from feature 7 (Poz-43120) from the same site did not provide final answers regarding the time period in which it functioned, because of discrepancy between relative chronology and radiocarbon dating results. Such an outcome may be interpreted as a recognition of another early medieval chronological phase of the settlement, but there is no archaeological evidence for this.

These results show, therefore, that in some cases, it is necessary to conduct further investigation or to select other materials from the same period for dating purposes, in order to verify the chronology of a given feature, e.g in case of Szczodrzykowo (Poz-43124) or Zielęcin (Poz-43120). In general, most of the obtained results allowed positive verification of the time intervals of functioning of the archaeological sites determined on the basis of relative chronology.

8. Acknowledgements

We sincerely thank prof. Tomsz Goslar, dr Justyna Czernik and Julia Kozik Maciejewska from the Poznań Radiocarbon Laboratory for the AMS dating, devoted time, and fruitful discussions; dr Michał Woszczyk for his geological suggestion.

We also would like to thank Agnieszka Tokarczuk-Różańska and dr Christopher Korten for practical help, linguistic comments and archaeological suggestions. Special thanks to our families: Michalscy, Kasprzak, Szczepaniak and Krzyszowscy.

The research was funded by the MNiSW grant no N N307 059437 and Iuventus Plus IP2010 027870.

9. References

Almgren O., (1923). Studien über nordeuropäischen Fibelformen der ersten nachristlichen Jahrhunderte mit Berücksichtigung der provinzialrömischen und südrussischen Formen, Leipzig

Ashmore PJ. (1999). Radiocarbon dating: avoiding errors by avoiding mixed samples. *Antiquity,* Vol. 73, No.279, pp.124–130, ISSN 0003 598X

Brock F., Higham T., Ditchfield P., Ramsey C. B. (2010). Current pretreatment methods for AMS radiocarbon dating at the Oxford Radiocarbon Accelerator Init (ORAU). *Radiocarbon,* Vol. 52, No.1, pp.103-112, ISSN 0033-8222

Bronk Ramsey C. 2010. OxCal Program v4.1. http://c14.arch.ox.ac.uk/embed.php?File=oxcal.html.

Brown Y.A., Nelson, D. S., Vogel, J. S. & Southton, J. R. (1988). Improved collagen extraction by modified Longin method. *Radiocarbon,* Vol. 30, pp.171-177, ISSN 0033-8222

Długosz J., (1874). *Historia – Joannis Dlugossi…Historiae Polonicae,* A. Przeździecki (ed.), Vol.2, Kraków, p. 80; *(in Polish)*

Kara M., Krąpiec M. (2000). Possibilities of dendrochronological dating of early medieval strongholds from the region of Wielkopolska, Dolny Śląsk and Małopolska, In: *Polish territories in the tenth century and their significance in the shaping of a new map of Europe* H. pp. 303-327, UNIVERSITAS, ISBN 83-7052-710-8, Kraków; *(in Polish)*

Kobusiewicz M. (Ed.). (2008). *Prehistory of Wielkopolska from the Stone Age to the Middle Ages.* Wyd. Instytutu Archeologii i Etnologii PAN, ISBN 978-83-89499-50-9, Poznań; *(in Polish)*

Łosiński W., Rogosz R. (1983). Rules of classification and the taxonomical scheme of pottery, In: *Szczecin in the Middle Ages. Castle Hill,* Cnotliwy E., Leciejewicz L., Łosiński W. pp. 202-226, Wrocław; *(in Polish)*

Łosiński W., Rogosz R. (1986). Rules of classification of early medieval pottery from Szczecin, In: Problems of early medieval ceramics chronology in Wester Pomerania, Warszawa 1986, pp. 51 - 61; *(in Polish)*

McFadgen BG. (1982). Dating New Zealand archaeology by radiocarbon. *New Zealand Journal of Science,* Vol.25, pp. 379–392. ISSN: 0028-8365

Nawrocka D., Czernik J. and Goslar T., (2009). ^{14}C dating of carbonate mortars from Polish and Israeli sites. *Radiocarbon* Vol.51, No.2, pp. 857-866, ISSN 0033-8222

Nawrocka Michalska D., Michczyńska D.J., Pazdur A., Czernik J., 2007.: Radiocarbon chronology of the ancient settlement on the Golan Heights. *Radiocarbon,* Vol. 49, No.2, pp. 625-637, ISSN 0033-8222

Rebollo N. R., Cohen-Ofri I., Popovitz-Biro R., Bar-Yosef O., Meignen L., Goldberg P., Weiner S., Boaretto E., 2008. Structural characterization of charcoal exposed to high and low pH: implications for ^{14}C sample preparation and charcoal preservation. *Radiocarbon,* Vol.50, No.2, pp. 289-307, ISSN 0033-8222

Reimer PJ, Baillie MGL, Bard E, Bayliss A, Beck JW, Blackwell PG, Bronk Ramsey C, Buck CE, Burr GS, Edwards RL, Friedrich M, Grootes PM, Guilderson TP, Hajdas I, Heaton TJ, Hogg AG, Hughen KA, Kaiser KF, Kromer B, McCormac FG, Manning SW, Reimer RW, Richards DA, Southon JR, Talamo S, Turney CSM, van der Plicht J, Weyhenmeyer CE. (2009). IntCal09 and Marine09 radiocarbon age calibration curves, 0–50,000 years cal BP. *Radiocarbon,* Vol.51, No.4, pp.1111–50, ISSN 0033-8222

Stankowski W. (1996). Introduction to the geology of the Cenozoic, with particular reference to the Polish territory. Adam Mickiewicz University, Poznań, ISBN 8323207798; *(in Polish)*

Stanley H. Ambrosea (1990). Preparation and characterization of Bone and Tooth Collagen for Isotopic Analysis. *Journal of Archaeological Science,* Vol.17, pp.431-451, ISSN: 0305-4403

Szczepaniak M., Nawrocka D. and Mrozek-Wysocka M. (2008). Applied geology in analytical characterization of stone materials from historical buildings. *Applied Physics A: Materials Science & Processing,* Vol. 90, No.1, pp. 89-95, ISSN: 0947-8396

van Klinken GJ. (1999). Bone collagen quality indicators for paleodietary and radiocarbon measurements. *Journal of Archaeological Sciences Vol.*26, No.6, pp. 687–95, ISSN: 0305-4403

Section 3

Luminescence and Radiocarbon Measurements

5

Geochronology of Soils and Landforms in Cultural Landscapes on Aeolian Sandy Substrates, Based on Radiocarbon and Optically Stimulated Luminescence Dating (Weert, SE-Netherlands)

J.M. van Mourik, A.C. Seijmonsbergen and B. Jansen
University of Amsterdam, Institute for Biodiversity and Ecosystem Dynamics (IBED), Netherlands

1. Introduction

The landscape of the study area (fig. 1,2) is underlain by coversand, deposited during the Late Glacial of the Weichselian. In the Preboreal, aeolian processes reduced soil formation (Stichting voor Bodemkaratering, 1972) and from the Preboreal to the Atlantic a deciduous climax forest developed (Janssen, 1974). The geomorphology was a coversand landscape, composed of ridges (umbric podzols), coversand plains (gleyic podzols), coversand depressions (histic podzols) and small valleys (gleysols). The area was used by hunting people during the Late Paleolithic and Mesolithic (Nies, 1999). Analysis of the urnfield 'Boshoverheide', indicated that the population increased during the Bronze Age between 1000 and 400 BC to a community of several hundreds of people, living from forest grazing, shifting cultivation and trade (Bloemers, 1988). The natural deciduous forests gradually degraded into heath land. The deforestation accelerated soil acidification and affected the hydrology, which is reflected in drying out of ridges and wetting of depressions, promoting the development of histosols and histic podzols. Sustainable productivity on chemically poor sandy substrates required application of organic fertilizers, composed of a mixture of organic litter with animal manure with a very low mineral compound (Van Mourik et al., 2011a), produced in shallow stables (Vera, 2011). The unit plaggic anthrosol on the soil map of 1950 AD identifies the land surface, which was used for plaggen agriculture. At least since 1000 AD, heath management was regulated by a series of rules that aimed to protect the valuable heat lands against degradation (Vera, 2011). During the 11th, 12th and 13th centuries there was an increasing demand for wood and clear cutting transformed the majority of the forests in driftsand landscapes (Vera, 2011). The exposed landscape was subjected to wind erosion and accumulation which endangered heath, arable land and even farmhouses. As a consequence, umbric podzols, the natural climax soil under deciduous forests on coversand, degraded into larger scale driftsand landscapes, characterized by deflation plains (gleyic arenosols) and complexes of inland dunes (haplic arenosols) (Van Mourik et al., 2011b). In such driftsand landscapes, the majority of the podzolic soils in coversand has been truncated by aeolian erosion. Only on scattered sheltered sites in the landscape, palaeopodzols were buried under mono or polycyclic driftsand deposits. They are now the valuable soil archives for palaeoecological research.

The city of Weert was founded at the end of the 13th century on a deforested topographic ridge of dry sandy soils, surrounded by swampy heath lands (Nies, 1999). Around 1300 AD the citizens ensured the supply of fresh water for the city moat by the creation of the 'Weerter Beek', a canal to connect the moat with a wetland area near the present Belgium border (Salmans and Tillemans, 1994). The topographical map of 1550 AD (fig.1) shows a deforested landscape surrounding the city, with distinct zones of arable land and heath; the natural forest had already completely been transferred into a cultural landscape.

During the 18th century, the population growth and regional economical activity stimulated the agricultural productivity. Farmers introduced the innovative 'deep stable' technique to increase the production of fertilizers (Vera, 2011). Additional to mowed biomass, farmers collected heath sods, including the top of the Ah horizon of the humus forms. This consequently promoted heath degradation and sand drifting, resulting in the extension of driftsand landscapes. During the 19th century, farmers tried to find alternative fertilizers and authorities initiated reforestation projects. The invention of chemical fertilizers at the end of the 19th century marked the end of the period of heath management and plaggen agriculture (Spek, 2004; Van Mourik at al., 2011b; Vera, 2011). The heath was not longer used for the harvesting of plaggic matter and new land management practices were introduced. Heath was reclaimed to new arable land or reforested with Scotch pine. During the 20th century the landscape dramatically changed again through a shift towards industrialization and bio-industry. Geomorphological features belonging to the historical sand drifting and plaggen agriculture survived in the landscape and are now included in the geological inheritance.

During recent decades the interest and need for restoration ecology and geoconservation has increased on global and regional scale (Bal et al., 2001). In the Netherlands, a national ecological master structure was designed to recover the ecological quality and biodiversity and in this context attention was paid to the preservation and restoration of driftsand habitats and landscapes (Koster, 2009, 2010).

Soil maps often serve as abiotical archives for ecosystem restoration management. However, soil classifications are normally based on actual diagnostic properties and therefore neglect relicts of former phases in soil and landscape development. Consequently, soil maps only show the distribution of recent soil types and are thus useless to fully understand the interaction of natural and human processes in time and space. To overcome this gap in knowledge of long term impact of human land use on the development of landforms and soils, the results of three innovative methods, applied to a selection of formerly investigated palaeosols, are presented in this paper. Firstly, the application of OSL dating on formerly investigated and ^{14}C dated palaeosols, to improve the geochronology of the phases in landscape evolution. Secondly the application of biomarker analysis to select the plant species, responsible for the production of organic carbon, stored in humic soil horizons. Finally it is shown how the complete package of palaeoecological information can be processed into soil maps of paleo-landscapes using a geographical information system.

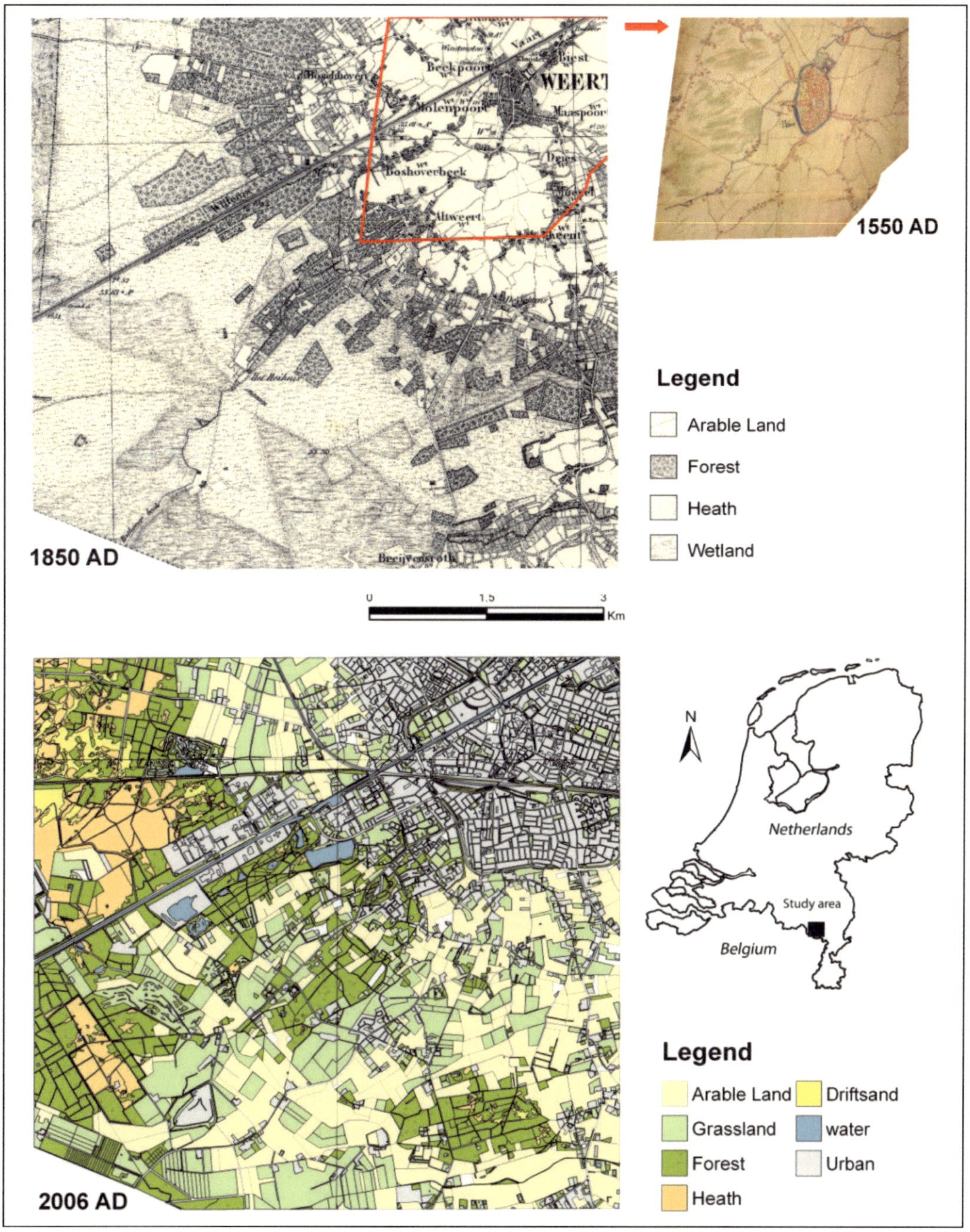

Fig. 1. Land use in the surroundings of Weert around 1550 AD (map of the City of Weert, surveyed between 1550 and 1570 by Jacob van Deventer), in 1850 AD (fragment of the topographical map scale 1:50,000, surveyed in 1837, published in 1863 by the Ministry of War) and 2006 AD (fragment of digital topographical map (Top10 vector), surveyed in 2006 AD).

2. Materials and methods

2.1 Profile selection

Since 1988 several pilot studies have been dedicated to the analysis of histosols, buried podzols and plaggic anthrosols around the city of Weert). Palynology, soil micromorphology and radiocarbon dating were the analytical tools to unlock the palaeoecological information from these valuable soil archives. For the reconstruction of the Late Holocene landscape evolution around the City of Weert, we selected several previously investigated key profiles. This selection comprises a histosol (Kruispeel), 3 buried histic podzols (IJzerenman, Tungelerwallen, Weerter Bergen), 3 buried (polycyclic) podzols (profiles Defensiedijk-1 and -2, Boshoverheide) and 3 plaggic anthrosols (profiles Tungelerakker, Dijkerakker and Valenakker).

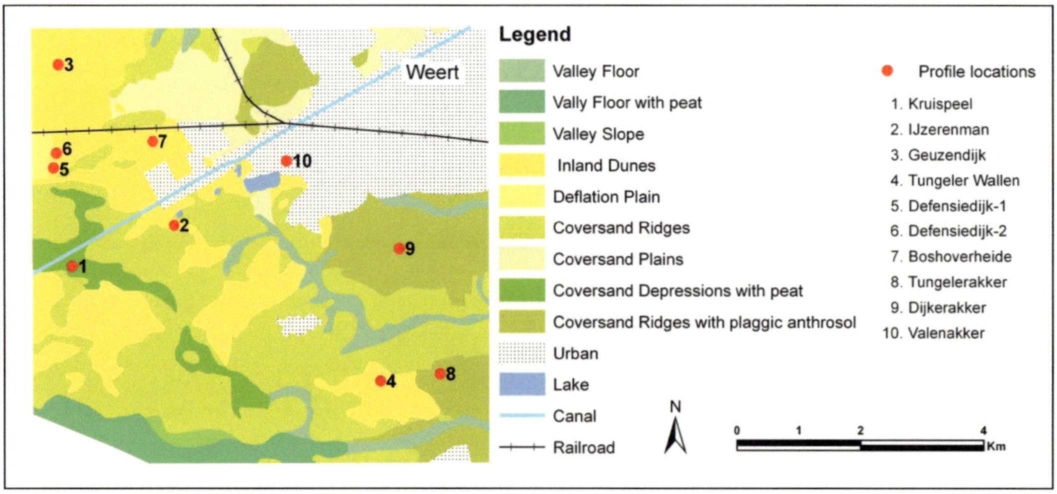

Fig. 2. Fragment of the geomorphological map, scale 1:50.000 with the profile locations.

2.2 Pollen analysis

Peat deposits are considered to contain syn-sedimentary pollen records and the diagrams reflect the characteristics of the local and regional vegetation development. Palaeosols are considered to contain a post-sedimentary pollen content as a result of bio-infiltration of pollen and the diagrams reflect ecological fingerprints of the soil ecological evolution (Van Mourik, 1999-b, 2001). Palynological reference of the Holocene vegetation development of SE-Netherlands, based on pollen analysis of extensive peat bogs in the Peel region have been published by Eshuis (1946) and Janssen (1974). The pollen zoning of the reference diagram is based on the zones of Firbas (1949). To understand the human impact on soils and landforms, local palaeoecological data was collected from a selection of palaeosols. From the profiles Kruispeel, IJzerenman and Weerter Bergen, samples were collected with an auger, vertical sampling distance 2.5 cm. From the other profiles, samples were collected in 10 ml tubes in profile pits, vertical sampling distance 2.5 cm. Pollen extractions were carried out using the tufa extraction method (Moore et al., 1991, p. 50). The exotic marker grain method was applied (Moore et al., 1991, p. 53) for estimation of the pollen densities in the mineral soils. For the identification of pollen grains the pollen key of Moore et al. (1991, p. 83-166) was applied. Pollen scores were based on the total pollen sum of arboreal and non arboreal plant species with the exception the aquatic species in pollen diagram Kruispeel. The pollen extractions were performed in the palynological laboratory IBED, University of Amsterdam. Pollen densities are represented in kilo grains / ml, or in the logarithmic value (log D)of the total amount of pollen grains / ml.

2.3 Soil micromorphology

Micromorphological observations are needed to assess the validity of soil ecological fingerprints, based on pollen spectra of drained soils. (Van Mourik, 2001). In thin sections of soil horizons, pollen grains are detectable but not determinable. The quality of the micro environment of pollen grains provide evidence of infiltration and preservation processes. Micromorphological observations are also relevant to understand the presence and distribution of soil organic carbon. Secondary soil processes can affect the quality of the original soil organic matrix and alter the organic chemistry composition and consequently the results of radiocarbon dating (Van Mourik et al., 2010). Undisturbed soil samples were collected in Kubiena boxes.

Undisturbed soil samples of buried humic horizons were collected in Kubiena boxes. Thin sections of undisturbed soil samples were produced, using the method of Jongerius and Heintzberger (1976) in the soil laboratory of IBED, University of Amsterdam. Thin sections were used to study the occurrence and preservation.

2.4 ^{14}C-dating

In traditional palynological research, radiocarbon dating was applied for absolute dating of pollen zones, as found in peat and limnic deposits. In palaeopedology, radiocarbon datings also have been used for dating purposes. The interpretation of radiocarbon datings of extracted soil organic carbon from soil samples is complicated. Due to the complexity of the provenance of soil organic carbon in humic soil horizons, conventional radiocarbon ages of bulk samples (BULK) are not very reliable. It is preferable to apply radiocarbon dating on extractions of fulvic acids (FUL), humic acids (HAC) and humin (HUM) (Goh and Molloy, 1978; Van Mourik et al., 1995, 2010). These fractions are based on extractability behavior. Fulvic acids are soluble in acid and in lye, the humic acids are insoluble in acid and soluble in lye and the humin fraction is insoluble in acid and in lye. The biological decomposition rate of FUL is relatively high; they migrate easily through weak acid soil profiles or leach completely. Therefore, they are unreliable for any dating purposes. The biological decomposition rate of HAC is medium high. Compared with FUL, they are immobile in the soil profiles and more reliable for dating purposes. The ^{14}C age of HAC is therefore expected to reflect the moment of burying of the soil. HUM (including pollen grains and charcoal) will accumulate in humic topsoil during an active period of soil development; therefore, ^{14}C ages of this fraction will definitely overestimate the age of burial. From these characteristics, we infer that the radiocarbon age difference between the HUM and HAC fractions of the same level will be greater during longer periods of active soil formation. To create a ^{14}C based geochronology, conventional ^{14}C dating was firstly applied on bulk samples of soil organic matter extracted from soil samples of the profiles Tungeler Wallen, Defensiedijk 1,2 and Boshoverheide; additional on FUL, HAC and HUM fractions of soil organic matter, extracted from samples of the profiles Defensiedijk 1 and 2 and Boshoverheide, and finally on HAC and HUM fractions of soil organic matter, extracted from samples of the profiles Kruispeel, Tungelruy, Dijkerakker,Weerter Bergen and IJzerenman. Radiocarbon datings have been performed in the *Centrum voor Isotopen Onderzoek* (CIO) of the University of Groningen, Netherlands (table 1). For calibration of the radiocarbon ages to calendar years, the program OxCal 4.1 (1 sigma confidence interval) was used.

2.5 Luminescence dating

For a general description of the OSL dating method is referred to Wintle (2008). Applications of OSL dating to plaggic deposits have been discussed earlier by Bokhorst et al. (2005) and Van Mourik et al. (2011b), while applications to polycyclic driftsand sequences were publishe in Van Mourik et al. (2010). The OSL samples were collected in standard pF-rings. Luminescence measurements used an automated Risø TL/OSL reader (DA 15) equipped with an internal Sr/Y beta source, and blue and IR diodes (Bøtter-Jensen et al., 2000). A single-aliquot regenerative dose (SAR) procedure was used for equivalent dose estimation (Murray and Wintle, 2003). OSL ages were calculated by dividing the sample burial dose by the dose rate and can be expressed in years relative to the year of sampling as well as in AD / BC. Quoted uncertainties contain all random and systematic errors in both the dose rate and burial dose assessment, and reflect the 1-sigma confidence interval. Luminescence dating of profile Defensiedijk 1 was performed at the Department of Geology and Soil Science, Laboratory of Mineralogy and Petrology (Luminescence Research Group), Ghent University (UG); of profile Dijkerakker at the Institute of Geography and Earth Sciences, University of Wales, Aberystwyth (UW); of profile Valenakker and Boshoverheide at the Netherlands Centre of Luminescence dating of the Delft University of Technology (NCL).

2.6 Biomarker analysis

Two pilots were chosen for the application of biomarker analysis. The first pilot was a polycyclic driftsand sequence. The research question was: Does the combination of pollen and biomarker analysis allow for a selection of the responsible plant species for the production of biomass and sequestration of soil organic carbon in buried humic horizons? We selected *n*-alkanes and *n*-alcohols with carbon numbers 20 - 36, which are exclusive to the epicuticular wax layers on leaves and roots of higher plants (Kolattukudy et al., 1976), as biomarkers of past vegetation. Recently, analysis of n-alkane and *n*-alcohol patterns preserved in Ecuadorian soils enabled a reconstruction of past vegetation dynamics in the area (Jansen et al., 2008). To explore the applicability of biomarker analysis for vegetation reconstructions in a wider range of soils we applied it as an additional proxy in a selected profile in the Weert setting: Defensiedijk-1. To this end the same A horizon samples were used as for the analysis of fossil pollen in the profile. In addition, leaves and roots from species expected to have been responsible for the dominant biomass input in the profile were sampled from the present day vegetation. These consisted of *Polytrichum piliferum*,

Cladonia rangiferina, Calluna vulgaris, Molinia caerulea, Corynephorus canescens, Deschampsia flexuosa, Pinus sylvestris, Betula pendula and *Quercus robur*. Approximately 0.1 g of each of the freeze-dried and ground vegetation and soil samples was extracted by Accelerated Solvent Extraction (ASE) using a Dionex 200 ASE extractor. The extraction temperature was 75°C and the extraction pressure 17 × 10^6 Pa, employing a heating phase of 5 min and a static extraction time of 20 min. Dichloromethane/methanol (DCM/MeOH) (93:7 v/v) was used as the extractant (Jansen et al., 2006a). The extracts were pre-treated and derivatized with BSTFA (*N,O*-bis(trimethylsilyl) trifluoroacetamide) containing 1% TMCS (trimethylchlorosilane) following a previously described protocol (Jansen et al., 2006b). Sample analysis took place on a ThermoQuest Trace GC 2000 gas chromatograph connected to a Finnigan Trace quadrupole mass spectrometer (MS), Separation took place by on-column injection of 1.0 µl of the derivatized extracts on a 30 m Rtx-5Sil MS column (Restek) with an internal diameter of 0.25 mm and film thickness of 0.1 µm, using He as a carrier gas. Temperature programming was: 50°C (hold 2 min); 40°C/min to 80°C (hold 2 min); 20°C/min to 130°C; 4°C/min to 350°C (hold 10 min). Subsequent MS detection in full scan mode used a mass-to-charge ratio (*m/z*) of 50-650 with a cycle time of 0.65 s and followed electron impact ionization (70 eV). The *n*-alkanes and *n*-alcohols were identified by their mass spectra and retention times and quantified using a deuterated internal standard (d_{42}-*n*-C_{20} alkane and d_{41}-*n*-C_{20} alcohol) (Jansen et al., 2006b). The concentrations of *n*-alkanes and *n*-alcohols with carbon numbers 20-36 in vegetation and soil samples were subsequently used as input for the VERHIB model that was specifically designed to translate such biomarker patterns into the most likely past vegetations patterns (Jansen et al. 2010). As required boundary conditions for the model we assumed a leaf biomarker vs. root biomarker input ratio in the soil of 1:10. For the tree species we assumed the root input to be equally distributed with depth, while for the grass and heath species we assumed all root input to have taken place within the first 36 cm, with 75% of that within the first 2 cm. The lichen and moss species were assumed to have given input only at the surface. The second pilot was a plaggic anthrosol. The research question was: Does the combination of pollen and biomarker analysis enable to detect the origin of the collected biomass for the preparation of plaggic manure? Three samples were analyzed from the plaggic horizons of profile Valenakker. Because of the limited number of horizons and the number of samples was too small to apply the VERHIB model.

2.7 Processing in ArcGIS

Three soil maps for the Weert area have been prepared in a digital environment in ArcGIS 10.0 (http://www.esri.com), following the method described by van Mourik et al. (2011) for a nearby area in the Netherlands. First, the soil map that represents the situation around 1950 and two historical maps, one of 1500 AD and one of 2000 BC. The 1950 soil unit boundaries are based on the scanned and digitized paper soil maps of the Netherlands (BKN50), whereas the two interpretative soil map boundaries are based on the results of the palaeoecological profile studies in this chapter and on available historical information of the vicinity of Weert. The digital 1:50.000 scale soil map of the Netherlands (BKN50 2006) is based on the analogue soil map of 1976 (Stichting voor Bodemkartering, 1976), which was mapped around 1968.

Soil code 1980	Soil name 1950 AD	Soil name 1500 AD	Soil name 2000 BC
Hd	Carbic podzol	Carbic podzol	Umbric podzol
Hn	Gleyic podzol	Gleyic podzol	Gleyic podzol
zEZ, EZg	Plaggic anthrosol	Plaggic podzols	Umbric podzols
pZg, pZn	Umbric arenic gleysol	Umbric arenic gleysol	Umbric arenic gleysol
aVz, zVz, Vz, Vp	Histosol	Histosols	Histic arenic gleysol
zWp	Histic podzol	Histic podzol	Gleyic podzol
zWz, vWz	Histic arenic gleysol	Histic arenic gleysol	Umbric arenic gleysol
Zd	Haplic arenosol	Umbric podzol	Umbric podzol
Zn	Gleyic arenosol	Gleyic podzol	Gleyic podzol
Ln	Siltic gleysol	Siltic gleysol	Siltic gleysol
Bebouwd	Urban	Urban	-

Table 1. Reclassification scheme used in GIS for the preparation of interpretive historical soil maps.

The translation of local (Dutch) soil type names in international labels is based on the World Reference Base (ISRIC/FAO, 2006). The vector-based soil map was clipped to the extent of the Weert study area by using a rectangular mask. The approximately 300 original legend categories were thus reduced to 17 legend units. These were further aggregated, based on similarities in soil texture properties, to 10 categories (table 1).

For the reconstruction of the soil maps of 1500AD and 2000 BC, detailed information on time development was used, which was obtained from the 10 key soil profiles described in detail in this chapter. This refers both to the age and to the palaeoecological information of buried soil horizons, which is indicative for renewed landscape dynamics, that is driven by either natural processes or man-induced interference. Crucial in this step is to define soil sequences for a time-span of approximately 4000 years, which is based on properties of the parent material, position in the landscape, local groundwater conditions over time and historical land cover and land use changes and, to a lesser extent, climate change. In the attribute table of the clipped digital soil geodatabase three additional columns were added for the 1950 AD, 1500 AD and the 2000 BC situations. Detailed, 5m resolution topographical information was used from the digital '*Algemeen Hoogtemodel Nederland*' (AHN5) which contains heights information expressed in cm. Land cover and land use data has been extracted from historical maps around 1550 AD, which shows an almost total deforested landscape with 'islands' of sand around the initial settlement of Weert and locations of former swampy areas. Visualization of the newly generated soil attributes leads to so-called 'interpretative historical soil maps' for the 1500 AC and 2000 BC situations. It was supposed that boundaries between the soil units did not significantly change over this time span. The soil boundaries that underlie the currently urbanized areas were reconstructed according to their likely fit with neighboring soil boundaries of the digital soil map of 2006, and by using historical map information.

3. Results and discussion

3.1 Palaeoecological information from a histosol and buried histic podzols

^{14}C Datings of (buried) histic horizons indicate the start of peat accumulation in coversand depressions around Weert between 1000 and 500 BC, due to wetting of depressions caused by deforestation of the surrounding higher coversand ridges. Similar datings were found in the study area Maashorst, 55 km north of Weert (Van Mourik et al., 2011b). In the more extensive depression of Kruispeel, the accumulation of peat continued during the Late Subboreal and Subatlantic; in small scale depressions gleyic podzols just transferred in histic podzols before they got buried under driftsand deposits around 1000 AD.

3.1.1 Profile Kruispeel, terric histosol (fig. 3-4; table 2;); pollen diagram first published in Van Mourik (1988)

Kruispeel used to be a shallow peat bog, situated in a depression in the coversand landscape (fig. 2). The formation of the histosols started in the Early Preboreal (*Pinus, Cyperaceaee*) with the deposition of humus gyttja, representing a shallow lake bottom soil, followed by a phase of terrestrialization with peat accumulation (*Betula, Salix, Artemisia, Helianthemum, Juniperus*). Compared with the Firbas pollen zoning of the Peel references (Janssen 1974), the composition of the pollen spectra is indicative for pollen zone IV. In the Boreal (zone V) the peat accumulation slowed down. The Atlantic (zones VI, VII) is absent. The accumulation of peat accelerated in the Late Subboreal (zone VIII) around 3300 BC (table 2) and continued in the Subatlantic (zones IX, X). The pollen spectra reflect a mix of species from the disappearing forest (*Corylus, Alnus, Quercus, Fagus*) and the emerging cultural landscape (*Ericaceae, Cerealia, Fagopyrum, Plantago*). *Cerealia* pollen is present in pollen spectra since the Bronze Age, *Fagopyrum* was introduced around 1350 AD (Leenders, 1987).

Diagram Kruispeel shows a clear expression of the impact of the vegetation development of the topographic higher surroundings on the hydrology of the depression. Deforestation on ridges (lower evapotranspiration, higher soil water infiltration) resulted in accumulation of gyttja or peat in the depression. Forested ridges (higher evapotranspiration and lower soil water infiltration) as a contrast resulted in a reduction of the accumulation rate or even in erosion by bio-oxidaten. The youngest spectra reflect the start of the period of reforestation (*Pinus*) since 1850 AD. The combination of reforestation and improve of the drainage by digging the Tungelroysche Beek at the end of the 19th century is responsible for stagnation of peat accumulation. Recently, wetland conditions and 'peel'ponds have been restored.

number	horizon	depth (cm)	fraction	^{14}C year BP	calibrated ^{14}C age
GrN 25419	2H	45-50	HUM	3440 ± 40	1738 - 1522 BC
GrN 25420	2H	45-50	HAC	3250 ± 40	1617 - 1437 BC

Table 2. ^{14}C datings of profile Kruispeel

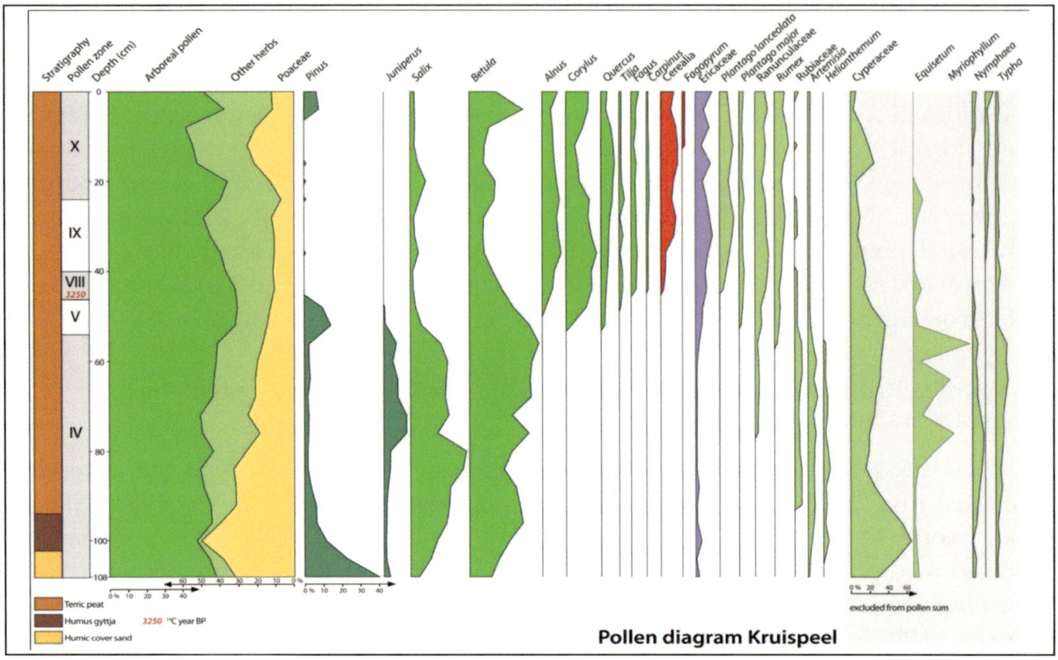

Fig. 3. Pollen diagram Kruispeel.

Fig. 4. Photo of the former peat bog Kruispeel, 2000 AD. The canal 'Tungelroysche Beek' was dug in the 19th century to improve drainage and land reclamation; the pond was created at the end of the 20th century as part of a nature development project to improve landscape quality and biodiversity.

3.1.2 Profile IJzeren Man, bi-cyclic haplic arenosol, overlying a histic podzol (fig. 5; table 3); pollen diagram first published in Van Mourik, 2000

Based on radiocarbon ages, the histic horizon of the palaeopodzol in coversand developed between 500 BC and 1000 AD. The post-sedimentary pollen spectra in the mineral horizon reflect decreasing scores of deciduous trees (*Corylus, Tilia, Quercus*), indicative for deforestation and high scores of *Ericaceae*, marking the extension of heath. Moist conditions during the development of the 3H or reflected by the scores of *Sphagnum*. Shortly after 1000

AD the histic podzol was buried by driftsand deposits. This age correlates with the period of forest clear cutting (Vera, 2011). The syn-sedimentary pollen spectra of the 2C are dominated by *Ericaceae*, reflecting heath dominance. After stabilization, a micropodzol (2AE) developed. It is difficult to establish an age, based on ^{14}C datings. The date of the HUM fraction probably overestimates the age and the date of the HAC fraction just indicates 'post Mediaeval'. The pollen spectra of the C are dominated by *Ericaceae* and *Poaceae*, reflecting a more degraded heath. After stabilization and reforestation *Larix* and *Pinus* pollen infiltrated, dominating the spectra of the actual micropodzol (3S).

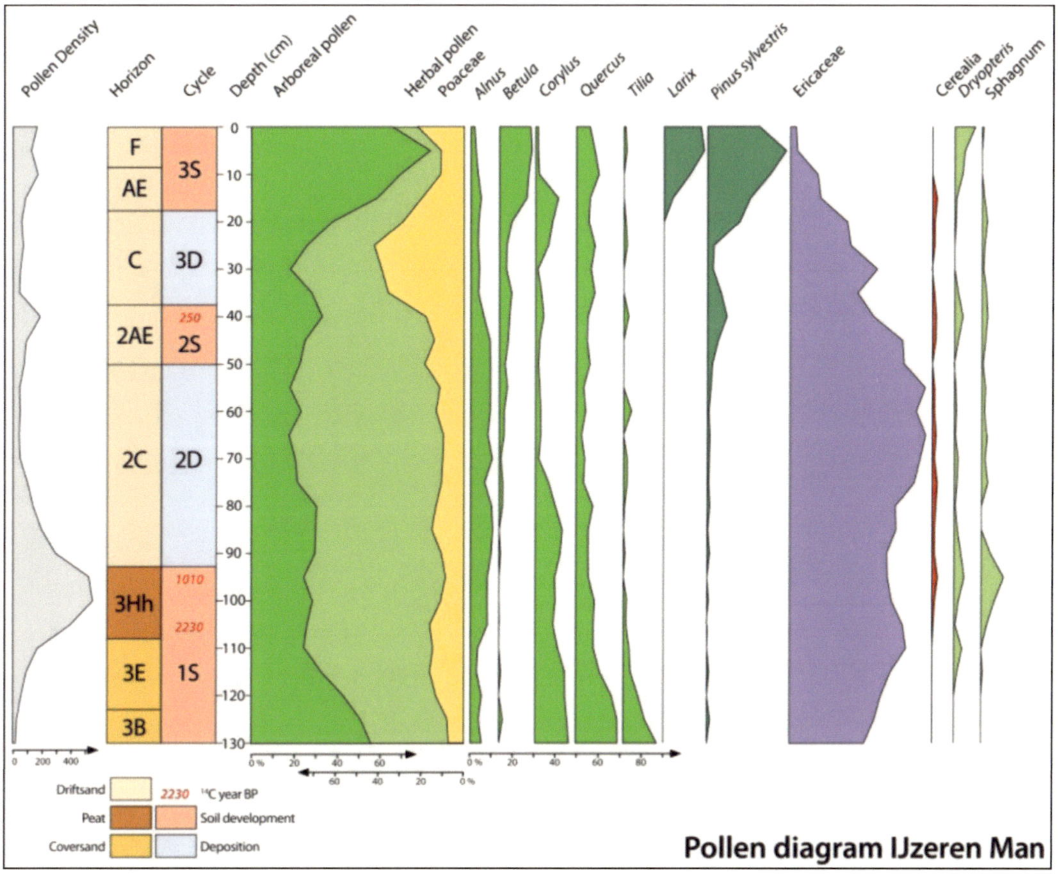

Fig. 5. Pollen diagram IJzeren man

Number	horizon	depth (cm)	fraction	^{14}C year BP	Calibrated ^{14}C ages
GrN 21245	2AE	40-24	HUM	1810 ± 90	18-417 AD
GrN 21245	2AE	40-42	HAC	250 ± 50	1482-1955 AD
GrN 23442	3AHh	90-91	HUM	1010 ±30	973-1152 AD
GrN 23443	3Hh	90-91	HAC	900 ± 40	1034-1215 AD
GrN 23444	3Hh	106-107	HUM	2230 ± 90	511-43 BC
GrN 23445	3Hh	106-107	HAC	2140 ± 35	358-56 BC

Table 3. ^{14}C datings of profile IJzerenman

3.1.3 Profile Tungeler Wallen, bi-cyclic haplic arenosol, overlying a histic podzol (fig. 6-7; Table 4); Pollen diagram first published in Van Mourik, 1988

Based on ^{14}C datings, the histic horizon developed between 300 BC and 1000 AD. It is unlikely that these ^{14}C datings are seriously affected by pollution by younger decomposition derivates from roots, considered the thickness of the driftsand deposits.

Number	horizon	depth (cm)	fraction	^{14}C year BP	Calibrated ^{14}C ages
GrN 14346	3Hh	135-140	BULK	945 ± 25	1027-1155 AD
GrN 24347	3Ah	160-165	BULK	2140 ± 30	353-56 BC

Table 4. ^{14}C datings of profile Tungeler Wallen

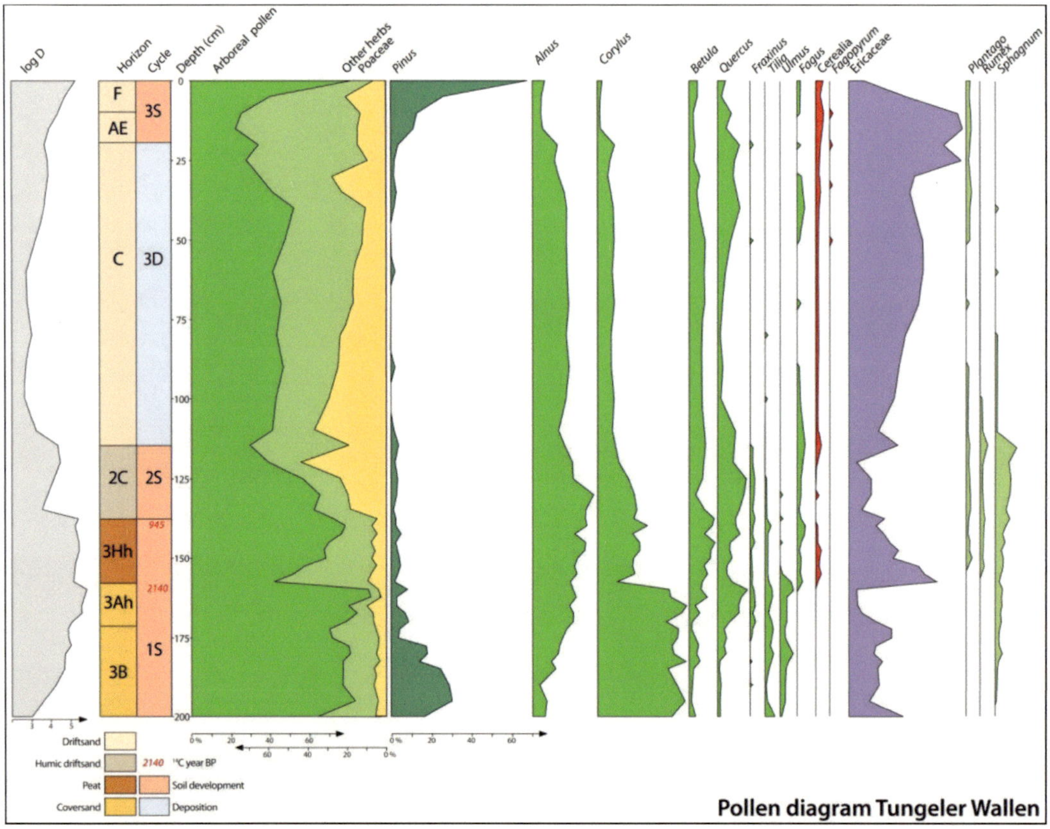

Fig. 6. Pollen diagram Tungeler Wallen.

The post-sedimentary pollen spectra of the palaeopodzol contain a mix of species from the Boreal forest (dominant in the 4B; *Corylus* and *Pinus*) and the Atlantic forest (dominant in the 3Ah; *Alnus, Quercus,Tilia, Ulmus*). The sharp fall of *Corylus* and rise of *Ericaceae* from the 3Ah to the 3Hh indicate a palynological hiatus in the record or even an erosion phase. Moist conditions during the development of the 3Hh are reflected by *Sphagnum*. During the field description two driftsand beds were distinguished. Based on the pollen density curve (log D), the lower bed was considered to represent the soil formation phase of the second cycle. Most probably, the 2AE horizon, expected above the 2C, has been eroded before the start of the third cycle. After stabilization and reforestation with Scotch pine, *Pinus* pollen could infiltrate and dominates the spectra the actual micropodzol (3D).

Fig. 7. Tungeler Wallen

3.1.4 Profile Weerter Bergen, tri-cyclic record; bi-cyclic haplic arenosol, overlying a histic podzol (fig.8-9; Table 5). Pollen diagram first published in Van Mourik, 1999a

Based on ^{14}C datings, the histic horizon developed between 400 BC and 600 AD. The pollen spectra of the palaeopodzol show, just as Tungeler Wallen and IJzerenman, high scores of elements of the former deciduous forest (*Corylus, Quercus, Tilia*), but also *Ericacea* show high scores, indicating heath in the surroundings. The development of the 3Hh took place between ± 200 BC and ± 700 AD. Special attention was paid in this profile on the determination of *Ericacae*, using the special pollen key of Moore et al. (1992) to distinguish *Erica tetralix*. The wetland vegetation is expressed by *Sphagnum, Myrica gale* and *Erica tetralix*. The pollen spectra of the 2C show a decrease of *Corylus* and an increase of *Ericacea*, indicating continuous extension of the heath. The ^{14}C datings of the 2 AE indicates a stable period around 800 AD under heath land conditions. After 800 AD, the micropodzol was buried under younger driftsand deposits. The pollen spectra of the C show very low percentages arboreal pollen and an increase of *Poaceae*, pointing to heath degradation. After stabilization and Pine plantation, *Pinus* pollen could infiltrate and dominate the spectra of the actual micropodzol.

Number	horizon	depth (cm)	fraction	^{14}C year BP	Calibrated ^{14}C ages
GrN 23436	2A	35-37	HUM	1150 ± 80	688-1020 AD
GrN 23437	2A	35-37	HAC	1165 ± 45	722-983 AD
GrN 23438	3Hh	67-68	HUM	1325 ± 50	610-809 AD
GrN 23439	3Hh	67-68	HAC	1380 ± 30	606-681 AD
GrN 23440	3Ah	78-80	HUM	2095 ± 80	361 BC-56 AD
GrN 23441	3Ah	78-80	HAC	2290 ±40	407-208 BC

Table 5. ^{14}C datings of profile Weerter Bergen

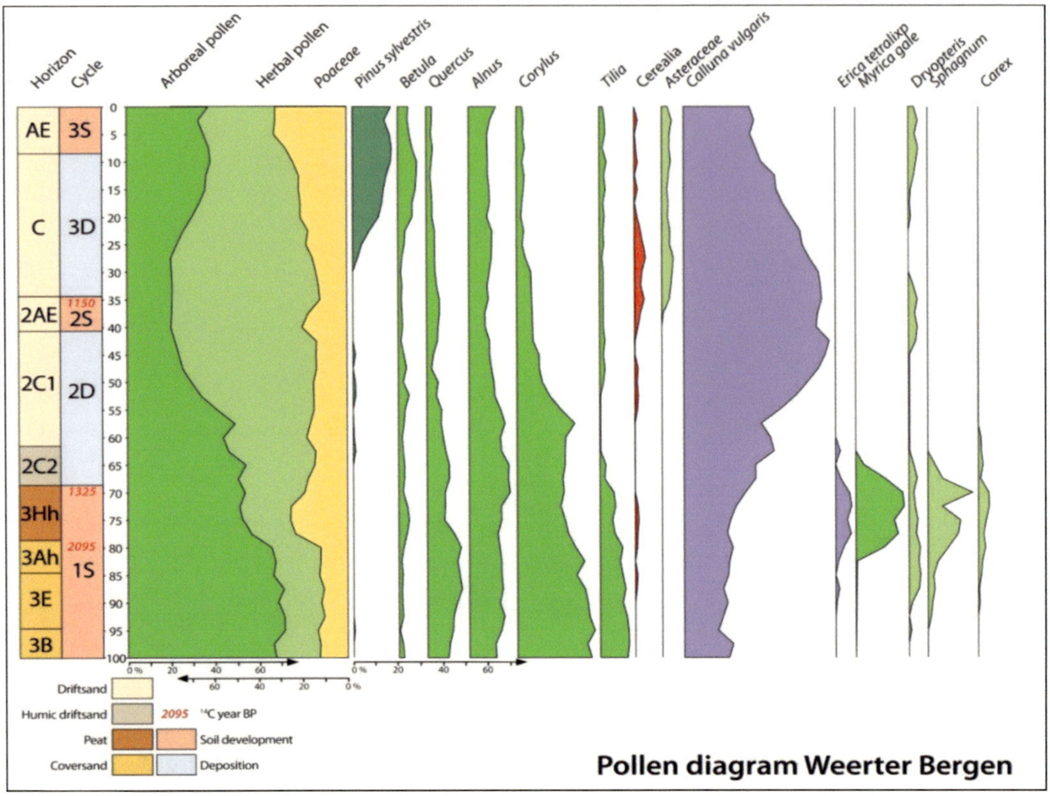

Fig. 8. Pollen diagram Weerter Bergen.

Fig. 9. Profile Weerter Bergen

3.2 Palaeoecological information from buried carbic podzols

Especially in well drained sandy landscapes, polycyclic sequences with buried podzols are unique parts of the soil archives. Pollen spectra provide information about the vegetation during stable and instable period. Application of biomarker analysis allow the selection of species, rooting during stable periods and responsible for the sequestration of soil organic carbon. The combination of ^{14}C and OSL datings indicates that the oldest palaeopodzols have been buried by Pre-Mediaeval small scale driftsand deposits, probably related to natural causes as hurricanes and forest fires or early shifting cultivation. Around 1000 AD a more extensive land surface with podzols got buried by larger scale driftsand deposits, probably related to the period of forest clear cutting. A generation of micropodzols, developed in stabilized driftsands, was buried by younger driftsand deposits after the 17th century, probably related to heath degradation due to the application of deep stable management.

3.2.1 Profile Defensiedijk 1, bi-cyclic haplic arenosol, overlying a bi-cyclic carbic podzol, pollen diagram first published in Van Mourik, 1988

3.2.1.1 Defensiedijk 1, sampled in 1986 (fig.10-15; table 6)

Application of radiocarbon dating in paleopedology is supposed to provide information for the establishment of the geochronology of the landscape evolution, but the interpretation of radiocarbon datings from paleosols is more complicated than in surveys of peat bogs and limnic deposits. This is caused by differential biological decomposition of soil organic litter, resulting in various fractions with different chemical compositions and turnover rates. For that reason, fractionated radiocarbon dating was introduced in paleopedology. Goh and Molloy (1978) investigated the suitability of radiocarbon datings of soil organic matter in quaternary geology and demonstrated the important role of the methods of extraction of organic matter. Ellis and Matthews (1984) established the differences in radiocarbon datings of FUL and HAC in palaeopodzols. The interpretations of the results of fractionated ^{14}C dating are heterogeneous. Mattheuws and Quentin (1983) selected the HAC fraction of the HF horizon for dating of a buried podzol in Norway. Hammond et al. (1991) established the importance of fractionated ^{14}C dating. In their research of peats and organic silts, FUL and HAC were considered as contaminates, leaching from podzolic environments. Dansgaard and Odgaard (2001) interpreted in their research of a buried podzol in Jutland the age of HUM, extracted from the B horizon as indicative for the start of podzolisation, the age of extracted FUL as indicative for the age of burial. Van Mourik et al. (1995) selected the HUM fraction of buried A horizons as indicative for the moment of burial, the differences in age between HUM and HAC fraction as indicative for a period of active soil formation and HUM accumulation and the FUL fraction as contaminated.

Profile Defensiedijk 1 is a palaeoecological record of four geomorphological cycles. Every cycle consist of an instable (D = aeolian deposition) and a stable (S = soil formation) period. During instable periods, syn-sedimentary pollen grains were incorporated in the sediment (low pollen densities), during stable periods, post-sedimentary pollen grains could infiltrate in the soil (high pollen densities). The pollen diagram (fig.10) provides information of the vegetation during instable (syn-sedimentary pollen) and stable (post-sedimentary pollen) periods (Van Mourik, 2001). The development of carbic podzol of the oldest cycle could continue from the Preboreal till the Late Subboreal. The post-sedimentary pollen spectra from the coversand deposit (cycle 1S) seems to reflect the presence of heath (*Ericaceae*) in a deforesting environment (*Corylus, Quercus*).

The post sedimentary pollen spectra from the oldest driftsand deposit (cycle 2D) show the maximal *Ericaceae* scores and the fall of *Corylus*, heath in a deforested cultural landscape. Podzolisation in this deposit could continue from ± 100 BC till 1000 AD. The syn-sedimentary scores of cycle 3D show an increase of *Poaceae* and a decrease of *Ericaceae*, reflecting heath degradation and sand drifting, the post sedimentary spectra of cycle 3S high scores of *Pinus*, reflecting soil formation under Pine forest. The syn-sedimentary spectra of the youngest cycle (4D) are characterized by dominance of *Poaceae* and a drop of *Pinus*, reflecting again degradation and sand drifting.

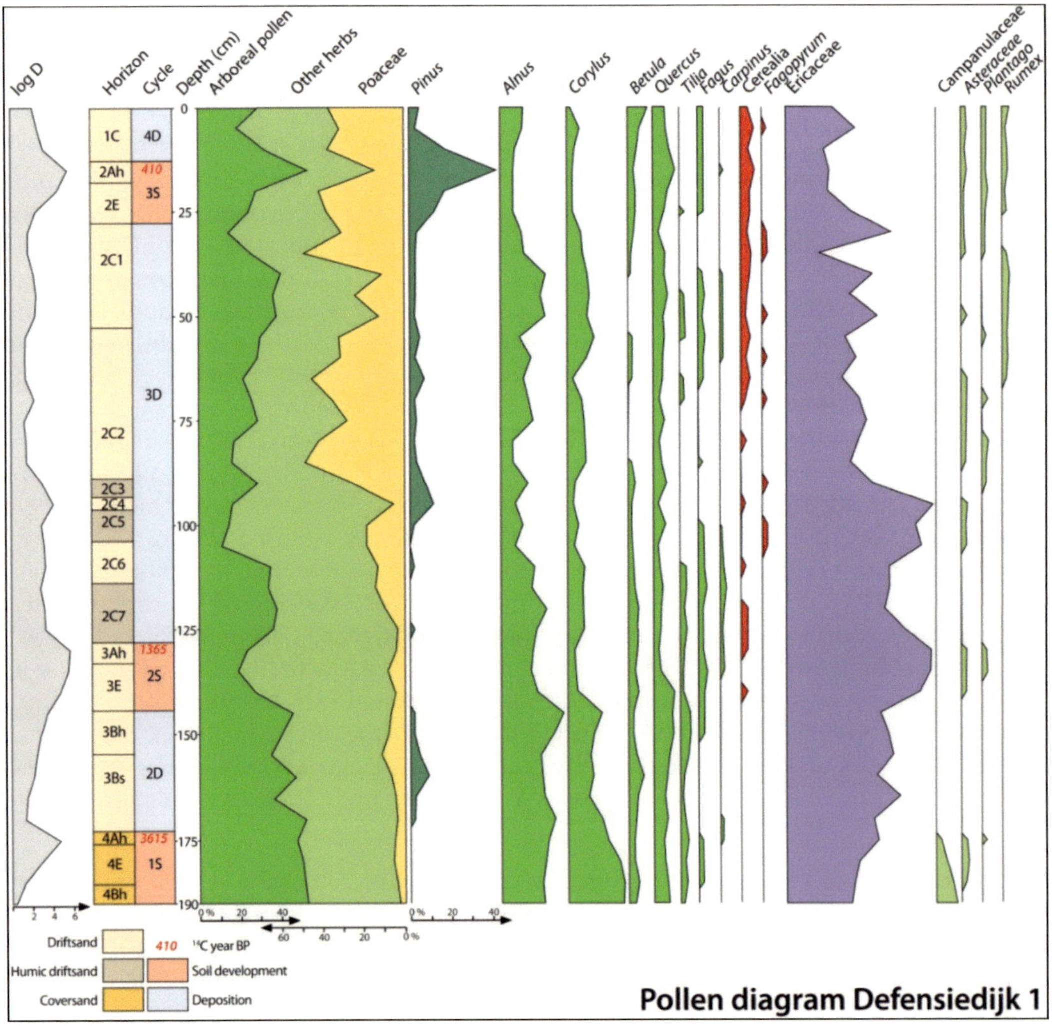

Fig. 10. Profile Defensiedijk 1.

Fig. 11. Defensiedijk 1, 1986

To establish the geochronology of the polycyclic sequence, ^{14}C dating was applied (table 6). The ^{14}C datings of BULK fractions were not satisfactorily for interpretation; especially the age of the 2A horizon was 'too old'. Soil micromorphological observations show the presence of transported organic aggregates and charcoal fragments in this horizon. That explains the 'old' age of the HUM fraction. The dataset sustains the conclusions (Van Mourik et al., 1995) that (1) ^{14}C ages of buried A horizons, based on HAC, are close to the moment of burial; (2) ^{14}C ages, based on HUM fractions are older than the moment of burial; (3) ^{14}C ages, based on FUL fractions are less reliable for understanding the geochronology. The ages, included in the pollen diagram, are the ^{14}C ages of the HAC fractions of the top of the buried A horizons.

number	horizon	depth (cm)	fraction	¹⁴C year BP	Calibrated ¹⁴C age
GrN - 14833	2Ah	025-027	HUM	3230 ± 110	1863-1219 BC
GrN - 14458	2Ah	025-027	HAC	410 ± 45	1423-1633 AD
GrN - 14759	2Ah	025-027	FUL	110 ± 120	1685-1928 AD
GrN - 12804	2Ah	025-027	BULK	1130 ± 45	778-999 AD
GrN - 14837	3Ah	127-129	HUM	1350 ± 50	602-775 AD
GrN - 14459	3Ah	127-129	HAC	1365 ± 25	520-687 AD
GrN - 14760	3Ah	127-129	FUL	850 ± 100	990-1380 AD
GrN - 14838	3Ah	129-131	HUM	1900 ± 110	166 BC-382 AD
GrN - 14694	3Ah	129-131	HAC	1675 ± 30	258-427 AD
GrN - 14774	3Ah	129-131	FUL	1200 ± 130	598-1149 AD
GrN - 12805	3Ah	127-131	BULK	1365 ± 25	620-687 AD
GrN - 14836	4Ah	173-174	HUM	4110 ± 90	2889-2475 BC
GrN - 14460	4Ah	173-174	HAC	3615 ± 35	2123-1887 BC
GrN - 14761	4Ah	173-174	FUL	2200 ± 170	736-123 BC
GrN – 14840	4Ah	175-176	HUM	4430 ± 165	3628-2638 BC
GrN - 14698	4Ah	175-176	HAC	3965 ± 40	2577-2344 BC
GrN - 14775	4Ah	175-176	FUL	3010 ± 140	1605-851 BC
GrN - 12808	4Ah	173-176	BULK	3615 ± 35	2123-1887 BC

Table 6. ^{14}C datings profile Defensiedijk 1 (1986)

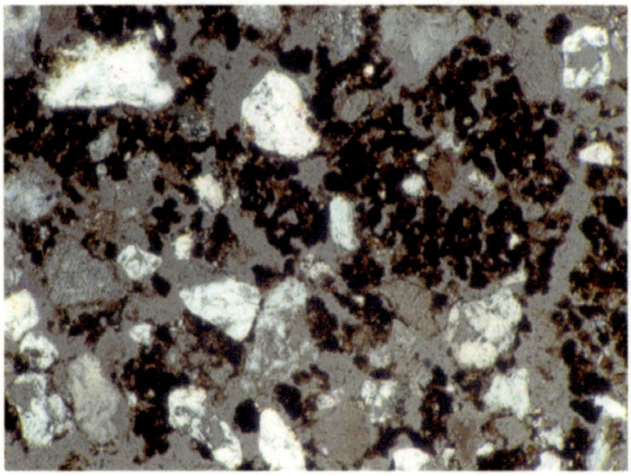

Fig. 12. Profile Defensiedijk 1. 2A horizon. Intertextic distributed organic aggregates with the intern fabric of fecal pallets; soil formation as result of litter decomposition by fungi and micro arthropods.

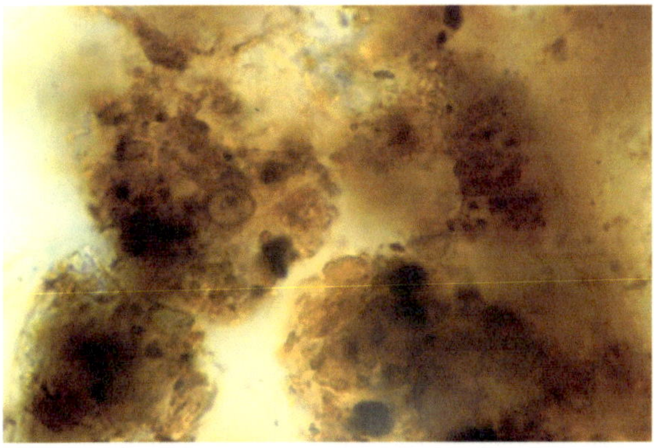

Fig. 13. Profile Defensiedijk 1. 2A horizon. Intern fabric of organic aggregates, showing the incorporation of small charcoal fragments and pollen grains.

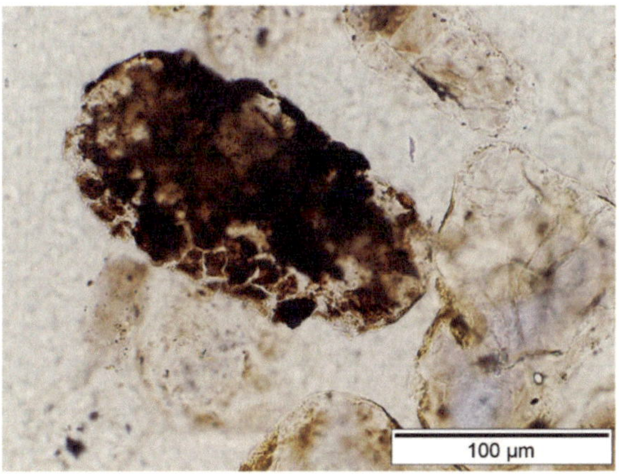

Fig. 14. profile Defensiedijk 1. 2C horizon. Rounded, transported organic aggregate, indicating sin-sedimentary contamination by older organic matter.

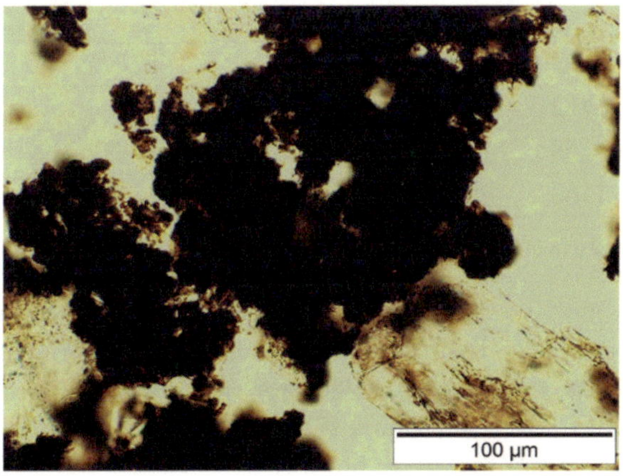

Fig. 15. Profile Defensiedijk 1. 4A horizon. Post-sedimentary, ageing intertextic distributed organic aggregates, indicating undisturbed soil structure.

3.2.1.2 Defensiedijk 1, sampled in 2002 (fig.16-17; table 7)

The geochronology, only based on ^{14}C dating, does not allow temporal separation during one cycle for sand drifting and soil development. Application of OSL dating can solve this problem. Due to aeolian erosion, it was impossible to resample on exactly the same site profile Defensiedijk 1. The horizontal distance between the profiles 1986 and 2002 was 8 meter. The thickness of the oldest driftsand formation was reduced to 15 cm. Defensiedijk 1 has been resampled in 2002 for OSL dating. Control samples were taken for ^{14}C dating (HAC fraction) of the 2AE, 3A and 4A horizons.

The geochronology can be improved (table 7) if ^{14}C and OSL datings are combined. The OSL age of cycle 1 is in line with the age of the coversand formation (Late Dryas). The ^{14}C control dating of the 4A is 'too young' in relation with the OSL age of the 3E horizon. Soil micromorphological observations indicate undisturbed patterns of intertextic modexi, but biomarker analysis points to significant rejuvenation of the organic matrix. The ^{14}C ages of the HUM fractions (table 3) of profile 1986 seem to fit better. The OSL age of the oldest driftsand formation is Neolithic. Maybe sand drifting was related with very early shifting cultivation or with a natural cause (hurricane, forest fire), resulting in aeolian erosion and deposition of small scale low inland dunes. The ^{14}C HAC ages of the 3Ah and 2AE fit rather good with the OSL age of the sequence of burying driftsand deposits.

number	horizon	depth (cm)	fraction	^{14}C year BP	Calibrated ^{14}C age	OSL age
W 49	1C1	45	Quartz			1920 AD ± 8
W 54	1C2	55	Quartz			1902 AD ± 10
W 48	1C3	70	Quartz			1912 AD ± 10
GrA 35143	2AE	105	HAC	0410 ± 35	1429-1627 AD	
W 44	2AE	105	Quartz			1652AD ± 30
W 3	2C1	110	Quartz			1412 AD ± 50
W 46	2C1	125	Quartz			1332AD ±100
W 14	2C2	137	Quartz			670 AD ± 60
GrA 35145	3Ah	145	HAC	1230 ± 35	687-885 AD	
W 21	3E	152	Quartz			2698 BC ± 400
GrA 35161	4Ah	158	HAC	2645 ± 40	897-778 BC	
W 24	4Bh	163	Quartz			7198 BC ± 800

Table 7. ^{14}C and OSL datings Defensiedijk 1 (2002)

The composition of pollen, precipitating on and infiltrating in a soil, is a mixture of pollen, dispersed by species rooting on the site and by species, present on distance. Consequently it is impossible to select the species, responsible for soil formation and humus sequestration from pollen spectra alone. Plant leaves are dispersed over much shorter distances by wind than pollen whereas plant root material normally enters a soil record in-situ, except for such cases where a large scale human induced deposition takes place e.g. through plaggen agriculture. As a result, in contrast to pollen records, biomarker records are expected to reflect better the local plant species responsible for soil formation and humus sequestration. For this purpose, Defensiedijk 1 was resampled again in 2008. Samples were taken from the mineral humic 1(A), 2Ae, 3Ah and 4Ah horizons and the humic driftsand layers 1C3, 1C5, 2C3 to compare pollen and biomarker spectra. Also a reference base was created for biomarkers of species, possibly involved in the carbon sequestration during soil formation (*Pinus, Betula, Quercus, Calluna, Molinia, Corynephorus, Polytrichum, Cladonia*)

Fig.16 summarizes the combined results of pollen and biomarker analyses from profile Defensiedijk 1, 2008. Micromorphologically, the intertextic modexal organic aggregates in buried Ah and Ah horizons seem undisturbed, but biomarkers indicate that the original tissue derived compounds can be overruled by younger root derived compounds. This observation is confirmed by a comparison of the pollen and biomarker results.

Fig. 16. Defensiedijk 1, 2002

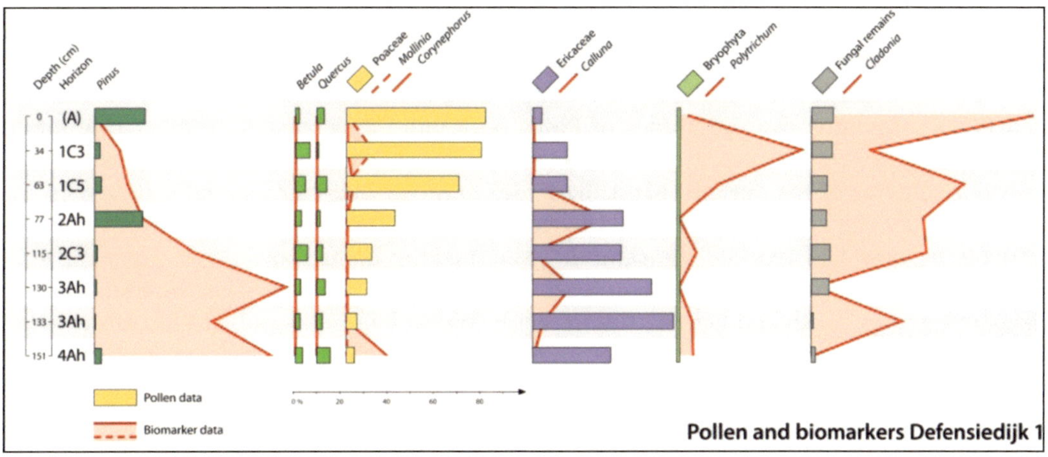

Fig. 17. Pollen and biomarkers profile Defensiedijk 1, 2008.

Fig. 18. Defensiedijk 1, 2008

The pollen spectrum of the 4Ah horizon is dominated by *Ericaceae, Corylus* and *Alnus* and micromorphological observations indicate an undisturbed soil matrix. Pollen grains can be extracted from organic aggregates with a modexal intern fabric and an intertextic distribution pattern. However, no biomarkers derived from *Ericaceae* or *Corylus* are present. Instead in the biomarker based reconstruction, *Pinus* and *Poaceae* are dominant. Pine trees were not introduced in the area until the 19th century and therefore are unlikely to represent the onsite vegetation at the time that the 4Ah horizon was at the surface, given the results of the dating of the horizon (Table 7). Instead, the dominance of Pinus biomarkers most likely represents 'contamination' of the soil organic carbon in this horizon with younger decomposition products of the roots of this deep rooting species. At the same time, the low abundance of *Ericaceae* in the biomarker based reconstruction most likely indicates that this species was absent at the site in significant numbers. Its abundance in the pollen records has been caused by windblown dispersal of pollen from surrounding areas. A cover dominated by grass and moss species as inferred by the biomarker reconstruction seems more likely.

The pollen spectrum of the 3A horizon is dominated by *Ericaceae*. The micromorphological structure is similar to the 4Ah, while the biomarker spectrum is now dominated by *Ericaceae* and *Pinus*. Assuming the abundance of Pinus once more to represent younger root input, heath would appear to have been the dominant vegetation at the site at the time that the 3Ah horizon was at the surface. Most likely the *Ericaceae* that were already present in the vicinity of the site earlier, as indicated by their presence in the pollen spectra of the 4Ah horizon, by now had reached the sampling site, making them show up in the more local biomarker based reconstruction .When the 2Ah horizon was at the surface, the site was most likely covered by heath and lichens, the former being present in the pollen spectra as well as the biomarker reconstruction, the latter showing up only in the biomarker reconstruction since it does not produce pollen. Only some fungal spores are present in the pollen extractions but they do not allow the identification of the fungi, associated with lichens that occur on driftsand and heath (*Cladena* and *Cladonia*)(Domsch and Gams, 1970; Aptroot and Van Herk,1994).

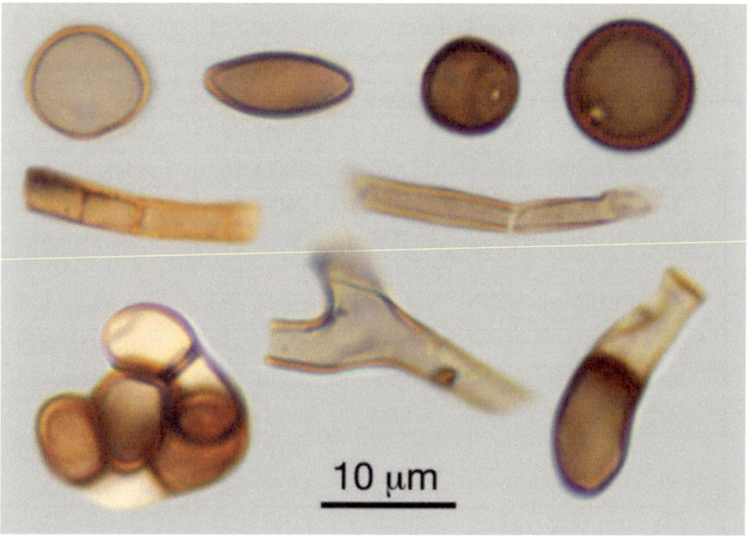

Fig. 19. Fungal remains in pollen spectra of the (A) horizon of profile Defensiedijk 1 (2008). These remains are also present in the other and humic horions

The ochric (A) horizon forms the present surface. The pollen spectrum is dominated by *Poacae*, the biomarker spectrum by lychens. The actual vegetation is dominated by *Cladonia* and *Polytrichum*. *Poaceae* are present in the surroundings, *Corynephorus* on dry inland dunes, *Molinia* in moist depressions. The biomarkers from grasses, specific for dry soil conditions (*Corynephorus canescens*, *Deschampsia flexuosa*) and for moist soil conditions (*Molinia caerulea*) are distinctive, in contrast to the pollen grains of these species. Palynologically, all driftsand layers of the youngest cycle are dominated by *Poaceae* and *Ericaceae*. Based on biomarkers, we can discriminate that the humus in the 1C5 stems from an eroded soil under a vegetation of lichens and dry grasses (probably *Deschamsia*), and in the 1C3 from an eroded site under lichens and Molinia. In both layers, markers of *Calluna* are absent in these spectra.

3.2.2 Profile Boshoverheide, bi-cyclic haplic arenosol, overlying a carbic podzol (fig.18-19; Table 8), pollen diagram first published in Van Mourik, 1988

The set of fractionated ^{14}C datings show similar features as the set of Defensiedijk 1. The ^{14}C ages of the oldest palaeosols (1S) are in line with the composition of the pollen spectra. Dominance of *Alnus* instead of *Corylus*, due to continuous infiltration of younger pollen grains and decay of fossil grains, points to a Middle Subatlantic palynological age; the development of the carbic podzol of the oldest cycle (1S) could continue from the Preboreal till ±1200 AD.

The burial of the palaeopodzol took place after 1200 AD. This is still in line with the period of forest clear cutting. Time, available for the development of the micropodzol (2S), de stable period between the depositions of S2 and S3, was (based on OSL datings) maximal 130. The radiocarbon ages of the carbon fractions, extracted from the 2Ah (with the exception of FUL) look 'too old', but the OSL age of the burial of the micropodzol is fits with the heat degradation after the introduction of deep stable management.

number	horizon	depth (cm)	fraction	^{14}C year BP	Calibrated ^{14}C age	OSL age
NCL 5106010	1C	027- 032	Quartz			1960 AD ± 2
GrN - 14834	2A	033-035	HUM	1710 ± 35	248-410 AD	
GrN - 14461	2A	033-035	HAC	390 ± 25	1414-1625 AD	
GrN - 14762	2A	033-035	FUL	240 ± 80	1470-1954 AD	
GrN - 12869	2A	033-035	BULK	265 ± 30	1516-1953 AD	
NCL 5106011	2C	035-040	Quartz			1830 AD ± 4
GrN - 14835	3A	125-127	HUM	830 ± 60	1042-1280 AD	
GrN - 14462	3A	125-127	HAC	615 ± 45	1286-1410 AD	
GrN - 14763	3A	125-127	FUL	490 ± 70	1297-1626 AD	
GrN - 14839	3A	127-129	HUM	1460 ± 90	403-768 AD	
GrN - 14702	3A	127-129	HAC	840 ± 45	1046-1274 AD	
GrN - 14773	3A	127-129	FUL	835 ± 55	1044-1277 AD	
GrN - 12870	3A	125-129	BULK	1190 ± 30	720-944 AD	

Table 8. ^{14}C and OSL datings of profile Boshoverheide

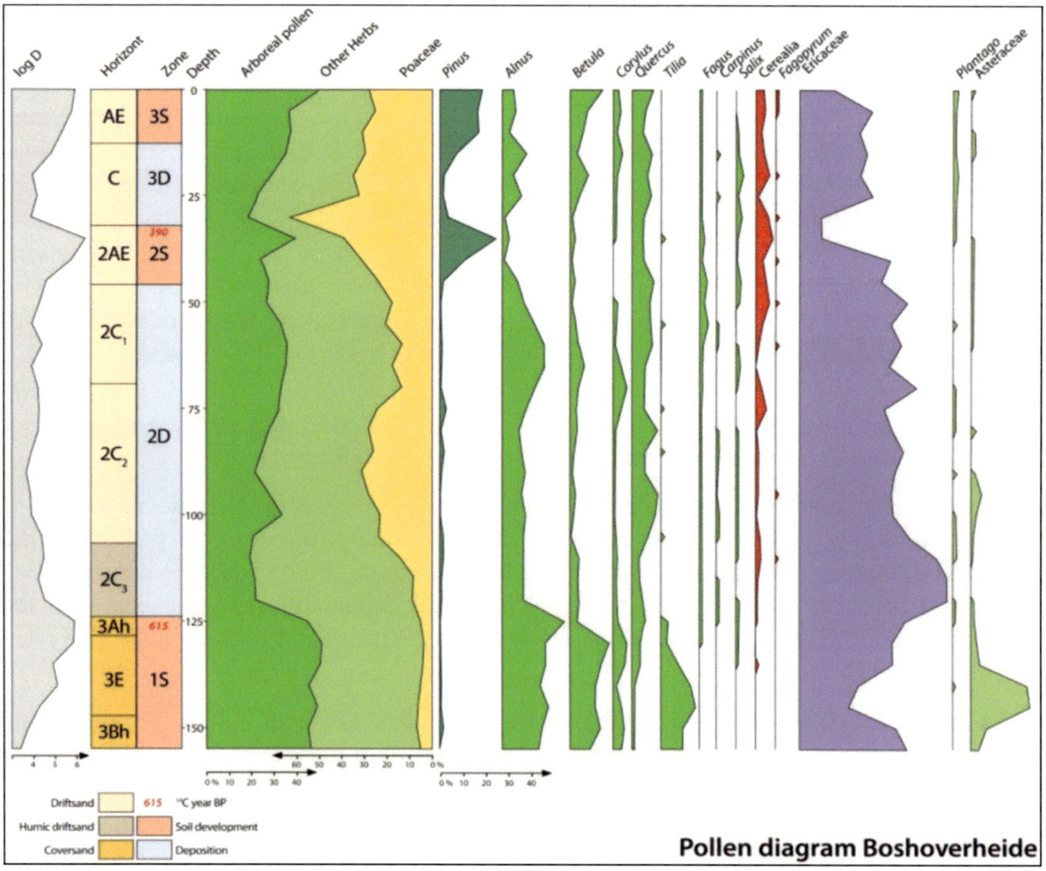

Fig. 20. Pollen diagram Boshoverheide.

Fig. 21. Profile site Boshoverheide.

3.2.3 Profile Defensiedijk 2, mono-cyclic haplic arenosol, overlying a bi-cyclic carbic podzol (fig.20-21, table 9), pollen diagram first published in Van Mourik, 1988

Profile Defensiedijk 2 is located 300 m NNE of Defensiedijk 1 near the border of a deflation plane and inland dunes. The palynological ages (post sedimentary pollen spectra with relatively high scores of *Corylus* in the oldest buried podzol, the decreasing scores of *Corylus* and the increasing scores of *Alnus* in the youngest buried podzol) and the ^{14}C ages (Bronze Age and Middle Ages) of the burial of the carbic podzols are comparable to Profile Defensiedijk 1. 3D is expressed by a 20 cm thick driftsand layer. The well preserved ^{14}C age of the 2S indicates that the driftsand deposits have been truncated. It is therefore impossible to distinguish between a mono or polycyclic driftsand package

number	horizon	depth (cm)	fraction	^{14}C year BP	Calibrated ^{14}C ages
GrN 13511	2A	35-37	BULK	1399 ± 35	582-675 AD
GrN 13512	3A	35-37	BULK	3140 ± 35	1498-1316 BC

Table 9. ^{14}C datings Profile Defensiedijk-2

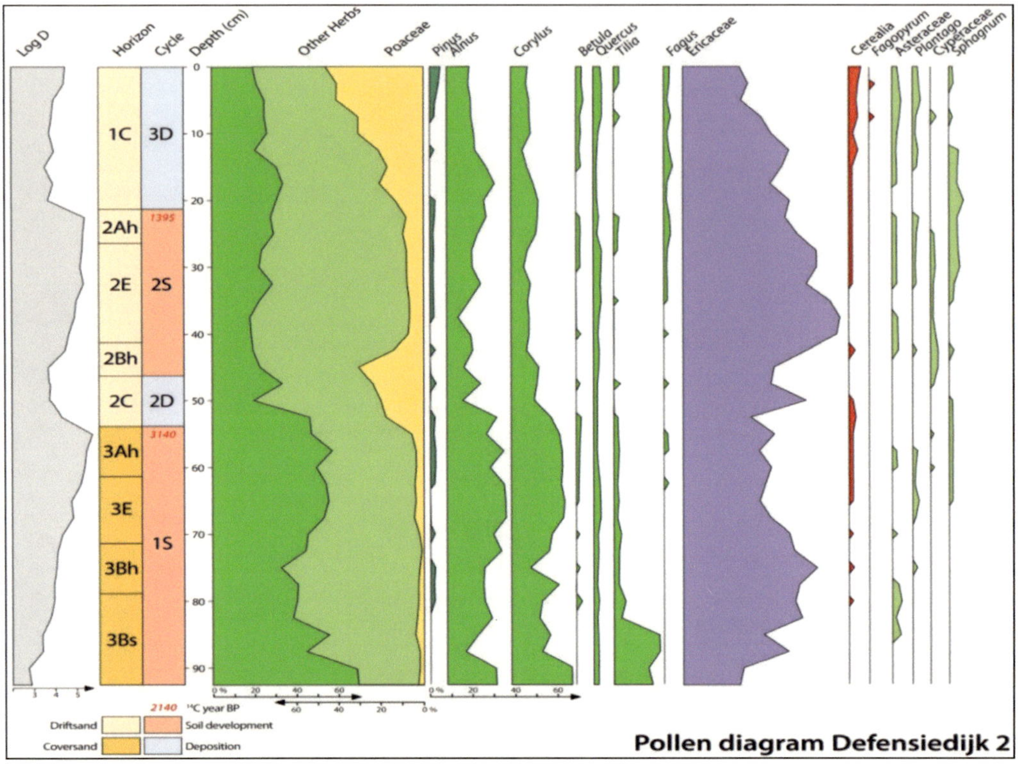

Fig. 22. Pollen diagram Defensiedijk 2.

Fig. 23. Profile site Defensiedijk 2.

3.3 Palaeoecological information from the ^{14}C and OSL dated plaggic anthrosols

The introduction and extension of plaggen agriculture on chemical poor sandy soils has long been controversial in historical geography and soil science (Spek, 2004; Van Mourik and Horsten, 1994, 1995). A complication was the interpretation of the validity of radiocarbon ^{14}C datings of organic extractions of plaggic deposits, due to the complexity of the composition of the soil organic matter (Van Mourik et al., 1995, 2010). This contradiction is very well illustrated by the observation that *Fagopyrum* pollen is already present in the lowest samples of plaggic deposits with radiocarbon age over 1000 year. Recently, the application of OSL dating in palaeopedology could clarify the genesis of plaggic anthrosols (Bokhorst et al., 2005, Van Mourik 2007; Van Mourik et al., 2011a). Based on radiocarbon datings agricultural land use is registered from approximately 1000 BC but OSL datings show that the accumulation of true plaggic deposits did not start before approximately 1500 AC. An older organic matrix is suspended in the younger mineral skeleton of plaggic deposits (Van Mourik et al., 2011a,b). The organic matrix consists of aggregates with an internal fabric, related to the internal fabric of modeled excrements of humus inhabiting soil fauna, consisting of organic plasma and micro particles, charcoal and pollen grains (fig.24-25, Van Mourik, 1999b, 2001). Consequently, palynological information and radiocarbon datings from plaggic anthrosols may contribute to the reconstruction of the evolution of vegetation and land use, but not to the absolute dating of plaggic deposits.

3.3.1 Profile Tungelerakker, plaggic anthrosol, overlying a ploughed (plaggic) podzol (fig. 22-23; table 10), Pollen diagram first published in Van Mourik, 1992

The first investigated plaggic anthrosol near Weert was profile Tungelerakker, situated 1050 m east of the Tungeler Wallen site, offering an optimal possibility to compare the pollen zoning in the buried histic podzol and the zoning in a plaggic anthrosol. The scores of arboreal pollen in the post-sedimentary pollen spectra from the ploughed palaeopodzol (S2) are low, the scores of *Ericaceae* are maximal and pollen of *Cerealia* is present. Based on the ^{14}C HUM and HAC ages, this indicates agricultural activity on the site in the period ± 230-770 AD. This fits with the ^{14}C BULK ages of the buried histic horizon in profile Tungeler Wallen, ± 300 BC-1100 AD.

number	horizon	depth (cm)	fraction	^{14}C year BP	Calibrated ^{14}C ages
GrN 17364	Aan	65	HAC	995 ± 60	896-1180 AD
GrN 17365	Aan	65	HUM	1390 ±120	405-936 AD
GrN 17366	Aan	105	HAC	1275 ± 75	639-948 AD
GrN 17367	Aan	105	HUM	1640 ± 80	234-595 AD
GrN 17368	2ABp	115	HAC	1370 ± 50	579-771 AD
GrN 17369	2ABp	115	HUM	1580 ± 110	230-660 AD

Table 10. ^{14}C datings of profile Tungelerakker

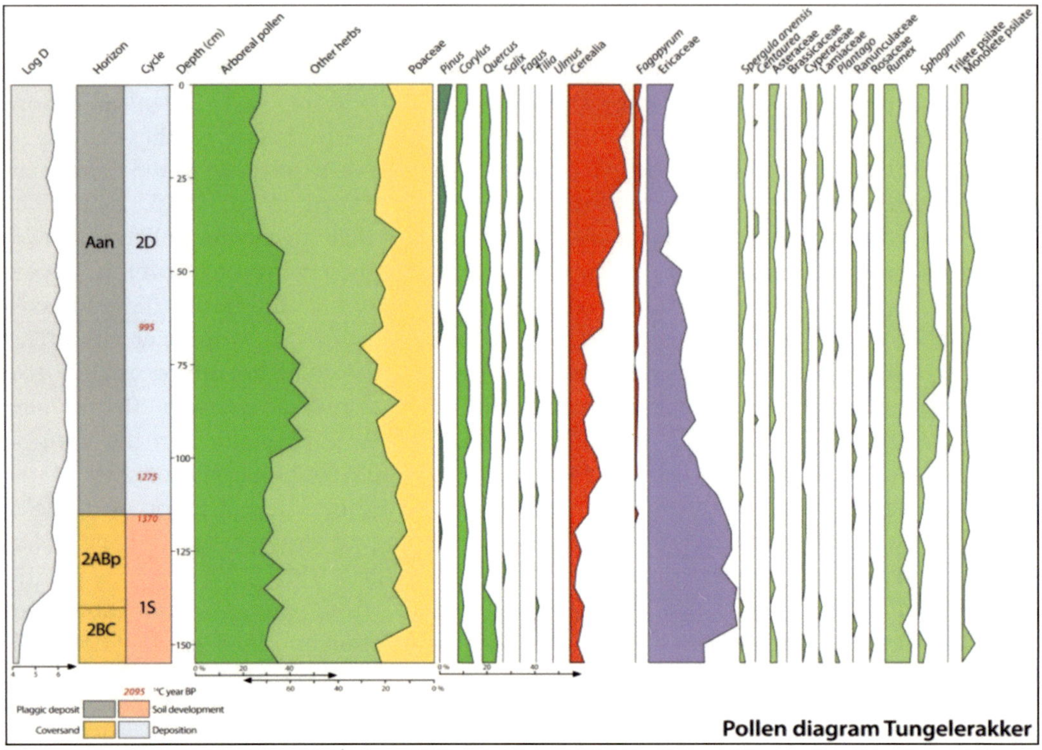

Fig. 24. Pollen diagram Tungelerakker.

Fig. 25. Profile Tungelerakker.

The ^{14}C ages of the plaggic deposits (2D) suggest plaggic deposition from ± 250-1400 AD. Application of OSL dating in more recently investigated plaggic deposits (Dijkerakker and Valenakker) indicate a contradiction between ^{14}C and OSL ages. It is proven that ^{14}C datings are valuable for the reconstruction of the agricultural development. Accurate dating of the plaggic deposits must be based on OLS ages. Palynologically, the Aan can be divided in two zones (2D1 and 2D2). 2D1 is characterized by high pollen densities, decreasing medium scores of *Ericaceae* and medium scores of *Cerealia*. The 2D2 is characterized by lower pollen densities, lower scores of *Ericaceae* and higher scores of *Cerealia*. This reflects probably the increasing crop production and the increasing content of mineral grains of the plaggic manure, related to the introduction of deep stable economy (Vera, 2011).

3.3.2 Profile Dijkerakker, plaggic anthrosol, overlying a (ploughed) carbic podzol (fig. 24-26, table 11), pollen diagram first published in Van Mourik and Horsten, 1994

The post sedimentary pollen spectra in the palaeopodzol (1S) show high scores of arboreal pollen (*Corylus, Quercus, Tilia*) and reflect a period of deforestation. Based on the ^{14}C datings of the 3Ah, the burial under driftsand (2D) took place around 3000 BP, rather similar to the oldest podzols in Defensiedijk-I and -2. The scores of *Cerealia* in the spectra of 1S and 2D point to agricultural activity and probably this (local) sand drifting was related to shifting cultivation. Farmers continued with agriculture on the 20 cm thick driftsand layer. Based on ^{14}C datings, the 2ABp fossilized around 1000 BP (839-1206 AD), close to the OSL age of the 2ABp of around 978 AD. No ^{14}C datings are available of the Aan, but the OSL age of the level 41-46 cm is 1607 AD. The set of datings supports the conclusion that agricultural land management was introduced before 1000 BC; The deposition of plaggic deposits started slow (30 cm between 1000 and 1600 AD) and accelerated during the last centuries of plaggen agriculture (45 cm between 1600 and 1900 Ad), probably related to the introduction of the deep stable agriculture in the 18th century.

OSL palaeodose measurements are based on 48 aliquots. The standard deviation but especially the distribution pattern of the values offers an excellent control on the validity of the dating (fig.31).

number	horizon	depth (cm)	fraction	^{14}C year BP	Calibrated ^{14}C ages	OSL age
A 74/1	Aan	46	Quartz			1607 AD ± 40
A 47/2	2ABp	76	Quartz			978 AD ± 208
GrN19488	2ABp	76	HAC	1000 ± 70	893-1206 AD	
GrN 19487	2ABp	76	HUM	1510 ± 40	433-637 AD	
GrN 19490	3Ap	90	HAC	3040 ± 80	1490-1049 BC	
GrN19489	3Ap	90	HUM	2830 ± 130	1385-794 BC	

Table 11. ^{14}C and OSL datings of profile Dijkerakker

The skewness of the histogram of sample 47/2 indicates that the calculated age is too high, caused by palaeodose overestimation. The true palaeodose cannot be accurately deducted because it is not known whether the skewness is caused by a large amount of partly bleached grains, by a smaller amount of much older unbleached grains or by a combination of these two factors. The aliquot measurements can be a mix of the age of mineral (2D) grains, never again bleached after sedimentation and grains, bleached at the surface after ploughing. Additional, bioturbation can be responsible for vertical transport of 'older' grains from below and younger grains from above.

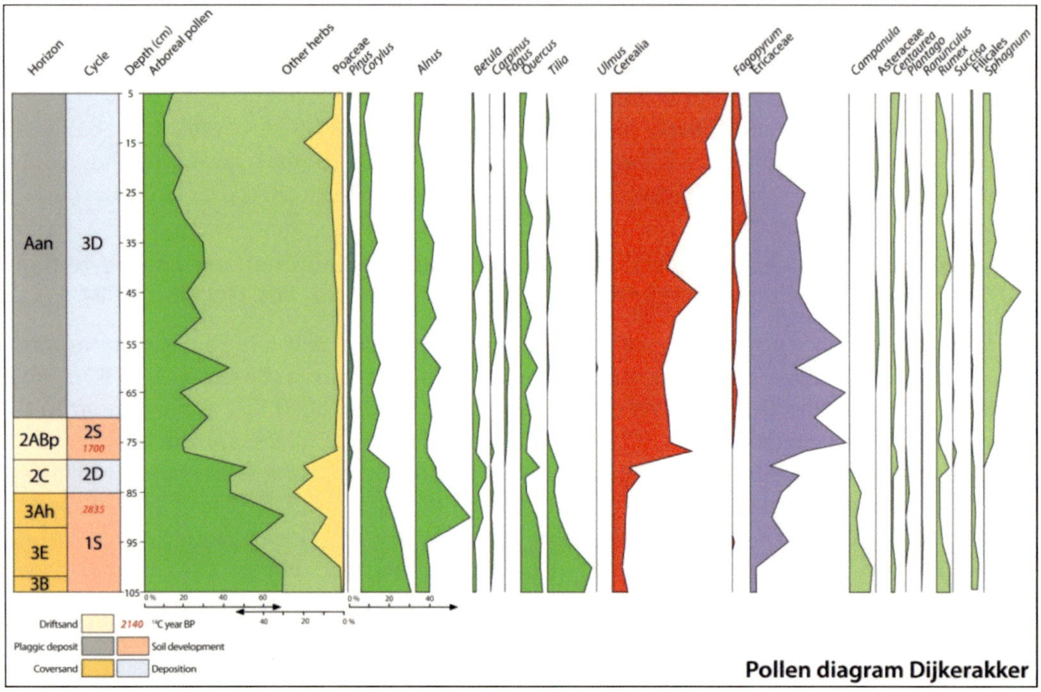

Fig. 26. pollendiagram Dijkerakker.

Fig. 27. Profile Dijkerakker.

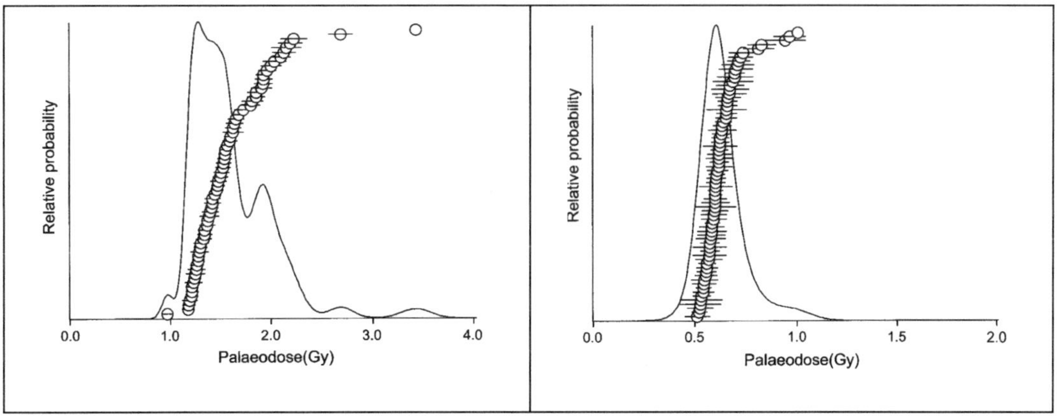

Fig. 28. Distribution of palaeodose values of the OSL datings of profile Dijkerakker. Left sample 47/2 (Dijkerakker 76 cm; 48 aliquots), right sample 47/1 (Dijkerakker 46 cm; 48 aliquots).

The histogram of sample 47/2 in fig. 5b indicates a reliable age, 394 ± 40 before 2001 (343 ± 40 BP). The majority of the mineral grains was completely bleached, promoted by the improvement of plough techniques, the increase of the mineral fraction of the plaggic manure and consequently the increase of the sedimentation rate.

3.3.3 Profile Valenakker, plaggic anthrosol, overlying a ploughed (plaggic) podzol (fig.27-31; table 12), pollen diagram first published in Van Mourik and Horsten, 1995

Profile Valenakker is well preserved in the urban environment of Weert on the sport field of a college and for more than 100 years never been ploughed. The post sedimentary pollen spectra in the BS show the low percentages of tree species as *Corylus* and *Quercus* of the Middle Subatlantic. The presence of *Cyperaceae* and *Sphagnum* indicates former most conditions of a gleysol or a gleyic podzol. Plough traces, together with the high scores of *Cerealia* in the 2Abp and even the 2B indicate a form of sedentary agriculture before the start of plaggen agriculture. Based on ^{14}C datings, plaggic deposition started around 500 AD. Based on OSL datings after 1500 AD.

GrN-nr	horizon	depth (cm)	fraction	^{14}C year BP	Calibrated ^{14}C ages	OSL age
NCL 5104001	Aan	20	Quartz			1755-1796 AD
NCL 5104002	Aan	40	Quartz			1605-1665 AD
GrN 21233	Aan	40	HAC	1000 ± 70	893-1206 AD	
GrN 21234	Aan	40	HUM	1270 ± 90	615-970 AD	
NCL 5104003	Aan	60	Quartz			1535-1595 AD
GrN 21235	Aan	60	HAC	1430 ± 80	429-768 AD	
GrN 21236	Aan	60	HUM	1340 ± 60	538-859 AD	

Table 12. ^{14}C and OSL datings of profile Valenakker

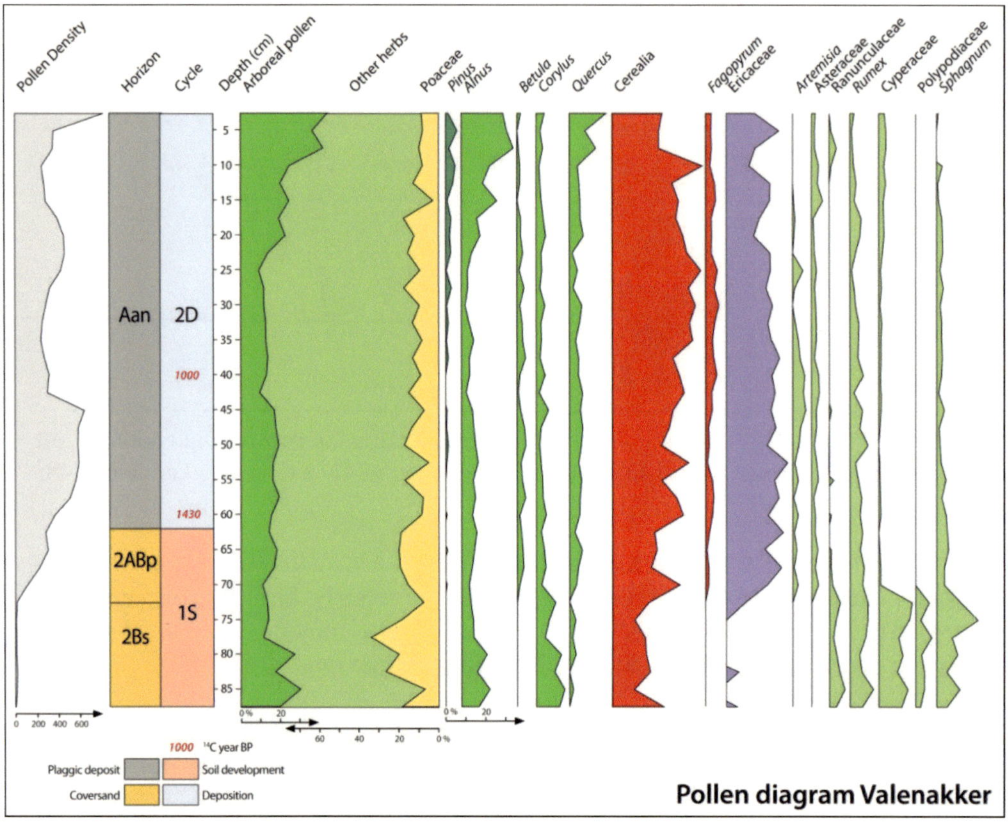

Fig. 29. Pollendiagram Valenakker.

Fig. 30. Profile Valenakker

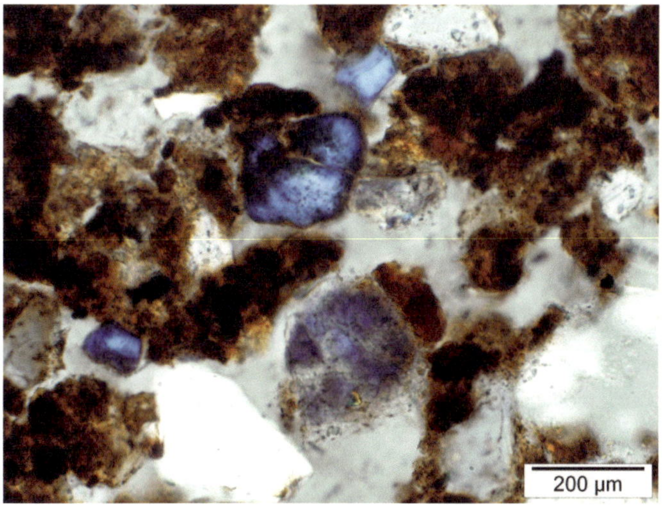

Fig. 31. Profile Valenakker Aan; soil skeleton of the Aan horizon, showing a partly intertextic, partly cutanic distribution of ageing organic aggregates.

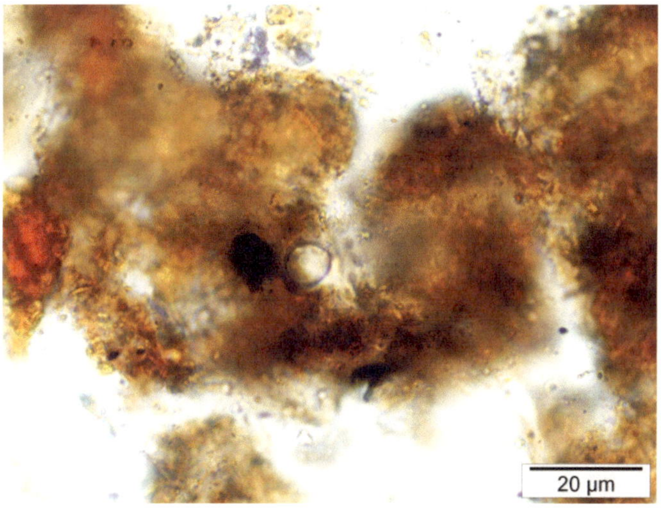

Fig. 32. Profile Valenakker Aan; intern fabric of ageing organic aggregate with charcoal fragments and pollen grain; the acid and micro-environment of excremental aggregates explains the pollen preservation under well drained soil conditions (Van Mourik, 2001).

Micromorphological observations show the complexity of soil organic matter in plaggic deposits. There are various sources of organic carbon as decomposing tissues of rooting plants, sods. The same is true for pollen spectra, a mix of the regular aeolian pollen influx and pollen, released from sods. The result is a plaggic deposit consisting of an older soil organic matrix, suspended in a younger mineral skeleton.

Traditionally, the origin of sods, used in plaggen agriculture, was reconstructed on the base of pollen spectra. The spectra of the Aan and the 2Abp show very low scores of arboreal trees but reasonable scores of *Ericaceae* and *Poaceae*. Ectorganic matter from forest soils is unlikely, *Ericaceae* pollen may indicate heat sods, *Poaceae* pollen on grassland sods, the combination on sods from degrading heath. It is very probably that during the formation of the 2Abp the farmers used mainly mowed heath shallow stables to produce manure for the arable land. During the formation of the Aan the collected sods with a higher mineral fraction to produce plaggic manure in deep stables (Vera, 2011). In fact, the origin of litter or sods cannot be satisfactorily detected with pollen diagrams. For that reason we applied biomarker analysis on two samples of the Aan and one sample of the 2Abp.

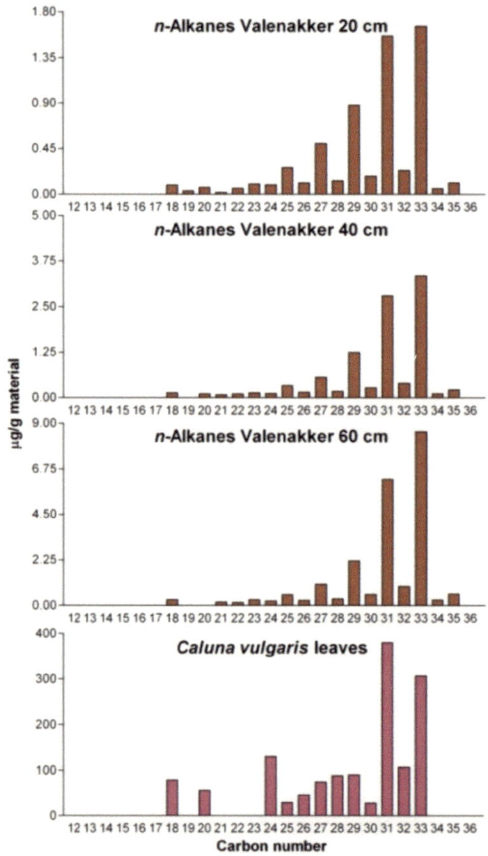

Fig. 33. n-alkane distribution patterns in samples of profile Valenakker and leaves of *Caluna vulgaris*.

In fig. 31 the n-alkane distribution patterns is shown in soil organic matter, extracted from soil samples, as well as in the leaves of *Caluna vulgaris*. While a complete biomarker reconstruction with the VERHIB model was not possible because of the limited samples numbers, a visual comparison of the patterns does yield some interesting information. The C_{31} and C_{33} *n*-alkanes dominated all three soil samples. Of the nine plant species considered (f 2.6) The leaves of *Caluna vulgaris* were the only to also have C_{31} and C_{33} n-alkanes as the dominant *n*-alkanes present. This, together with the uniformity of the *n*-alkane patterns at all depths sampled, is consistent with the history of plaggen agriculture on the profile and points to a dominant use of heath sods therein. It is also consistent with the importance of

Ericaceae in the pollen spectra. At the same time, visual assessment of the n-alcohol patterns (data not shown) does not show a clear link to a single plant species. In this respect the limited number of samples with respect to the soil profile and the plant species considered are a limiting factor in these exploratory analyses. Nevertheless, the results have great potential for the use of biomarkers to help reconstruct organic matter input in plaggen soils in combination with other proxies.

4. Reconstruction of historical soil maps

4.1 Soil map around 2000 BC

The surface parent material exists of coversands which were deposited during Late Glacial and Preboreal time. Initial soil formation started without human interference and continued until Late Subboreal time. As a consequence, the soil map shows an undisturbed pattern of climax soils: umbric and gleyic podzols on coversand ridges, Siltic, umbric and histic gleysols in depressions and valley's. From the Bronze Age on, the effect of human land use on soil forming processes increases. The deciduous forest degraded partially into *Calluna* heath as the result of forest grazing, small scale wood cutting and shifting cultivation. In a further phase, deforestation accelerated and shifting cultivation was replaced by sedentary forms of agriculture during the Iron Age. The soil map reconstructed for 2000 BC therefore shows a soil distribution of only few soil types which is strongly adapted to local coversand topography.

4.2 Soil map around 1500 AC

People learned to collect ectorganic and mowed biomass to produce in the last remains of the forest and on the heath. Plaggen agriculture was probably locally introduced in the Early Middle Ages (Spek, 2004) and became the regular land use system on mineral poor sandy soils in the medieval period. As an effect, the removal of ectorganic matter triggered soil acidification, and replacement of umbric podzols by carbic podzols. Another effect in the landscape was the disturbance of surface and soil water hydrology. Less water was intercepted by forests, which enhanced soil water infiltration. In lower terrain, in-between coversand ridges such as small valleys and local depressions, umbric gleysols developed into histosols. The intense application of plaggic manure on arable land prevented extensive acidification, so that plaggic podzols became a common soil types. The soil map reconstructed for 1500AC shows a much more fragmented pattern as a consequence of this early human interference in the previous natural soil conditions.

4.3 Soil map around 1950 AC

An increasing population demanded intensification of food production. During the 11th, 12th and 13th century clear cutting of the forested areas created an environmental catastrophe: extensive sand drifting (Vera, 2011). Deflation plains with gleyic arenosols and inland dune complexes with haplic arenosols were the results. After the introduction of the deep stable economy in the 18th century, farmers collected additional to mowed biomass more and more heath sod, including the endorganic humic mineral horizon. The mineral fraction of the plaggic manure increased. Sod digging resulted in a first local degradation of the Calluna heath, and over time regional effects were noticeable, that even prevented the regeneration

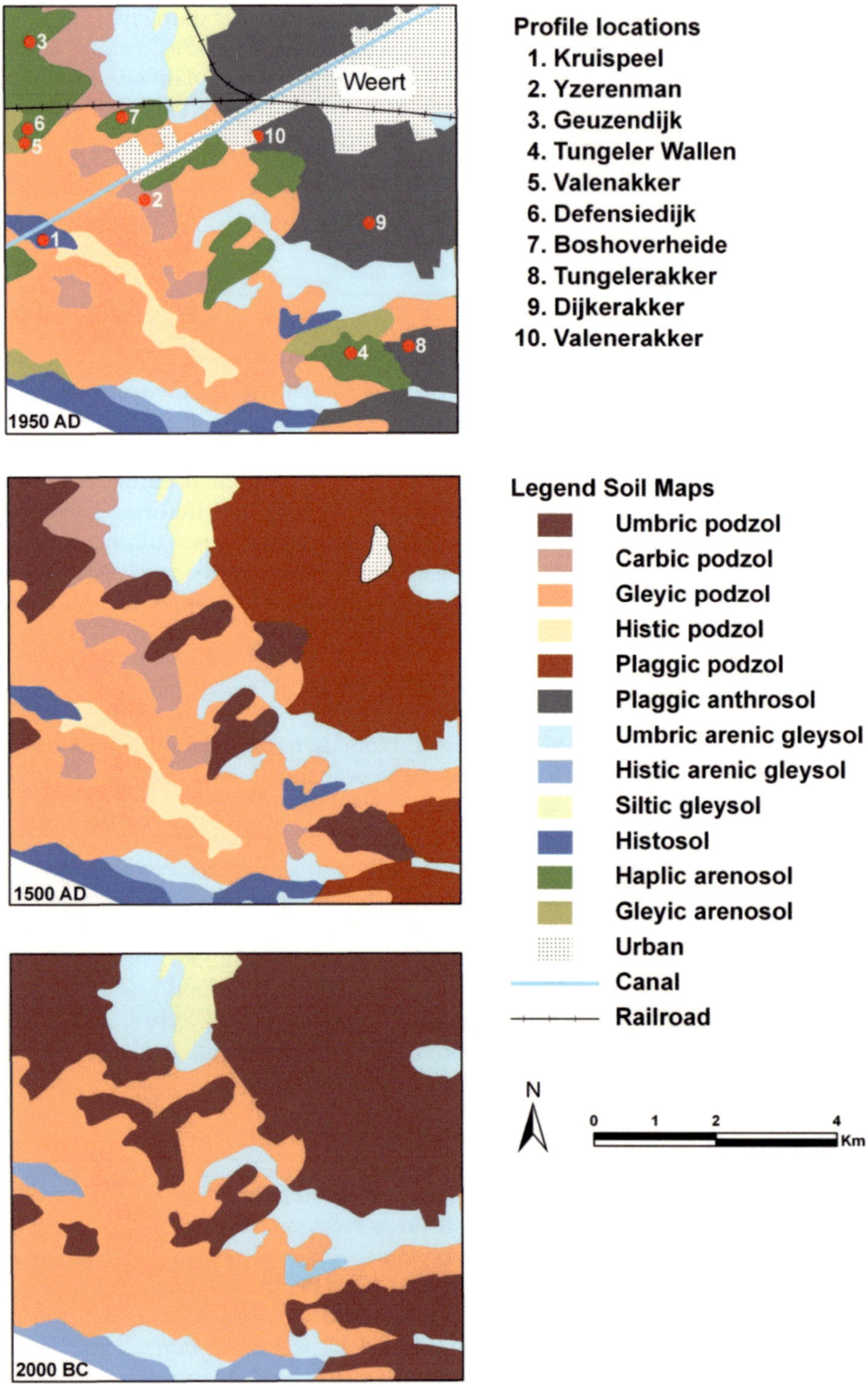

Fig. 34. The evolution of the soil geography

of the vegetation cover. In parts of the Late Glacial coversand landscape this triggered the extension of the typical Late Holocene driftsands. The consequence for soil formation was that plaggic podzols were replaced by plaggic anthrosols, mainly as the result of accumulation of the mineral compound of the plaggic manure. Degradation of the Calluna heath initiated the degradation of carbic podzols to xeromorphic and hydromorphic arenosols. The invention of chemical fertilizers around 1900 AD took away the need to collect sod manure for crop production. The information which is nowadays preserved in many polygenetic soil profiles proves that the resulting soil patterns are strongly controlled by human interference. The legend of the soil map of 1950 AD is based on the Dutch Soil Map, sheet 57-Escale 1:50,000 (Stichting voor Bodemkartering, 1972). The Dutch soil units are translated to the system of the World Reference Base (ISRIC/FAO, 2006).

4.4 Recent developments

Recent land management is further affecting soil development, which will continue to control the transformation of existing soil patterns. The former Calluna heath is partly reclaimed for arable land and partly turned into forest (dominated by Scotch pine). There is also a trend towards stabilization of driftsand areas by invasion and succession of vegetation. The reclamation of the Calluna Heath into arable land changed carbic podzols into anthric podzols. Under Scotch pine, mormoder humus forms developed (Sevink and De Waal, 2010) and haplic arenosols develop to albic arenosols. Peat accumulation was interrupted in the depressions, because soil water infiltration in the forest stands decreased and due to bio oxidation, histosols shift towards umbric gleysol.

5. Conclusions

Paleosols can be considered as important geo-ecological records, but due to the complexity of soil organic carbon, extracted from buried humic soil horizons, an accurate geochronology of such records cannot be based only on ^{14}C datings. The combination of OSL and radiocarbon datings enables the correlation of paleo-ecological data derived from histosols and mineral paleosols and improves the geochronology of the evolution of polycyclic soils and landforms in aeolian sands and plaggic deposits.

In palaeopodzols contradictions appear between OSL and ^{14}C ages. Soil micromorphological observations and the results of biomarker analysis show that the composition of soil organic carbon in buried humic horizons can be affected by secondary soil formation (buried podzols) and sedimentation (micropodzols).

In plaggic deposits radiocarbon datings reflect the development of an older organic matrix, suspended in a younger mineral skeleton. ^{14}C datings of the organic matrix are relevant for the reconstruction of the agricultural history, OSL datings of the mineral skeleton are relevant for the destination of the age of plaggic deposits.

Multi method analyses of polycyclic soil profiles provide the detailed knowledge which is necessary to fully understand time development of soil patterns in areas which are strongly affected by human land use. The combination of traditional soil survey techniques (soil classification, soil mapping), pollen analyses, micromorphology and soil dating techniques (^{14}C, OSL) makes it possible to date major changes in geo-ecological evolution.

Spatial reconstruction of former soils geographical patterns is possible in GIS through a back analysis of soil patterns based on reclassification of current soil types and soil boundaries.

Exploratory application of biomarker analysis in the form of preserved concentration patterns of C_{20} - C_{36} n-alkanes and n-alcohols showed to be a valuable additional method for the analysis of polycyclic soil profiles. In combination with fossil pollen analysis, it can help to distinguish former regional vegetation patterns from past local vegetation cover at a study site. It also has potential to yield information about the origin of plant derived organic matter in plaggen agriculture systems. However, application in this field of study is still in its infancy and further development of this technique is needed to uncover the full potential of the method.

6. References

Aptroot, A and Van Herk, K., 1994. Veldgids Korstmossen. KNVV Uitgeverij, Utrecht, Netherlands.

Bal, D., Beije, H.M., Fellinger, M., Haveman, R., Van Opstal, A.J.F.M., Van Zadelhoff, F.J., 2001. Handboek Natuurdoeltypen. Experisecentrum LNV, Wageningen, Netherlands. (with english summary).

Bloemers, J.H.F., 1988. Het urnenveld uit de Late Bronstijd en Vroege IJzertijd op de Boshover Heide bij Weert. In: Landschap in beweging; Ontwikkeling en bewoning van een stuifzandgebied in de Kempen, pp. 59-137. Nederlandse Geografische Studies 74, UvA/KNAG, Amsterdam.

Bokhorst,M.P., G.A.T Duller, G.A.T, Van Mourik, J.M., 2005; Optically Stimulated Luminescence Dating of a fimic anthrosol in the Southern Netherlands. Journal of Archaeological Science 2005, 547-553.

Bøtter-Jensen, L., Bulur, E., Duller, G.A.T., Murray, A.S., 2000. Advances in luminescence instrument systems. Radiation Measurements 32, 57-73.

Domsch. K.H. and Gams, W., 1970. Pilze aus Agraböden. Fischer Verlag, Stuttgart, Germany.

Ellis, S. and Matthews, A.J., 1984. Pedogenetic implications of a ^{14}C-dated paleopopodzolic soil at Haugabreen, Southern Norway. Arctic and Alpine research, vol. 16 - 1, pp. 77 - 91.

Eshuis, H.J. 1946. Palynologisch en stratigrafisch onderzoek van de Peelvenen. Schotanus and Jens publishers, Utrecht, Nederlands.

Firbas, F., 1949. Spät- und nacheiszeitliche Waldgeschichte Mitteleuropas nördlich der Alpen. Band 1: Allgemeine Waldgeschichte. Gustav Fischer Verlag, Jena, Germany.

Goh, K.M. and Molloy, B.P.J., 1978. Radiocarbon dating of paleosols using soil organic matter components. Journal of soil Science 29, p. 567-573.

Hammond, A.P., Goh, K. M., Tonkin, J. and Manning, M.R., 1991. Chemical pretreatments for improving the radiocarbon datings of peats and organic silts in a gley podzol environment: Grahams Terrace, North Westland. New Zealand Journal of Geology and Geophysics, Volume 34, Issue 2, 1991.

ISRIC-FAO, 2006. World Reference Base for Soil Recourses 2006. World soil resources reports 103.

Jansen, B., Nierop, K. G. J., Kotte, M. C., De Voogt, P., and Verstraten, J. M., 2006a. The application of Accelerated Solvent Extraction (ASE) to extract lipid biomarkers from soils. Applied Geochemistry 21, 1006-1015.

Jansen, B., Nierop, K. G. J., Hageman, J. A., Cleef, A. M., and Verstraten, J. M., 2006b. The straight-chain lipid biomarker composition of plant species responsible for the dominant biomass production along two altitudinal transects in the Ecuadorian Andes. Organic Geochemistry 37, 1514-1536.

Jansen, B. Haussmann, N.S., Tonneijck, F.H., De Voogt, W.P. and Verstraten, J.M., 2008. Characteristic straight-chain lipid ratios as a quick method to assess past forest - páramo transisitons in the Ecuadorian Andes, Palaeogeography, Palaeoclimatology, Palaeoecology, 262: 129-139.

Jansen, B., Van Loon, E. E., Hooghiemstra, H., and Verstraten, J. M., 2010. Improved reconstruction of palaeo-environments through unravelling of preserved vegetation biomarker patterns. Palaeogeography, Palaeoclimatology, Palaeoecology 285, 119-130.

Janssen, C.R., 1974. Verkenningen in de palynologie. Oosthoek, Scheltema & Holkema publishers, Utrecht, Netherlands.

Jongerius, A., Heintzberger, G., 1976. Methods in soil micromorphology. Soil Survey Papers 10, Wageningen, Netherlands.

Kolattukudy, P. E., Croteau, R., and Buckner, J. S., 1976. Biochemistry of plant waxes. In: Kolattukudy, P. E. (Ed.), Chemistry and biochemistry of natural waxes. Elsevier, Amsterdam.

Koster, E.A., 2009. The European Aeolian Sand Belt: Geoconcervation of driftsand landscapes. Geoheritage 1, 93-1120.

Koster, E.A., 2010. Origin and development of Late Holocene driftsands. In: Fanta, J., Siepel, H. (eds.) 2010. Inland driftsand landscapes.. KNVV Publishing, Zeist, Netherlands, 25-48.

Leenders, K.A.H.W., 1987. De boekweitcultuur in historisch perspectief. Geografisch Tijdschrift 21, p. 213-222. (with English summary).

Matthews, J.A. and Quentin Dresser, P., 1983. Intensive ^{14}C dating of a buried palaeosol horizon. Geologiska Foereningan i Stockholm. Volume 105, Issue 1, 1983.

Moore, P.D., Webb, J.A., Collinson, M.E., 1991. Pollen Analyses. Blackwell Scientific Publications, Oxford.

Murray, A.S. and Wintle, A.G., 2003. The single aliquot regenerative dose protocol: potential for improvements in reliability, Radiation Measurement 37, 377–381.

Nies, F., 1999. Weert, het verleden van een stad. Van Buuren BV, publishers, Weert, Netherlands.

Salmans, J.P.F. and Tillemans, H.L.A., 1994. Boshoverbeek, kroniek van een historische driehoek. Werkgroep Boshoverbeek, Weert, Netherlands.

Sevink, J. and De Waal, R.W., 2010. Soil and humus development in driftsands. In: Fanta, J., Siepel, H. (eds.) 2010. Inland driftsand landscapes.. KNVV Publishing, Zeist, Netherlands, 107-134.

Spek, T., 2004. Het Drentse esdorpenlandschap, een historisch geografische studie. Matrijs, Utrecht, Chapter 5. (with english summary)

Stichting voor Bodemkartering (1972). Bodemkaart van Nederland blad 57-oost met toelichting, 's-Hertogenbosch. Pudoc, Wageningen, Netherlands.

Van Mourik, J.M., 1988. Landschap in Beweging, ontwikkeling en bewoning van een stuifzandgebied in de Kempen, pp. 5-42. NGS 74, UvA/KNAG, Amsterdam.

Van Mourik, J.M., 1992. Het ontstaan van de Tungeler Wallen. Weerter Jaarboek 1993, 88-102.
Van Mourik, J.M., 1999a. Het aardkundig erfgoed van de Weerter Bergen; ontstaan en vergaan van een heideven. Weerterjaarboek 2000, 103-113.
Van Mourik, J.M., 1999b. The use of micromorphology in soil pollen analysis. Catena 35, 239-257.
Van Mourik, J.M., 2000. De toekomst van de IJzeren Man 2; fysisch geografisch onderzoek rondom de ontgronding van de CZW. Weerterjaarboek 2001, p. 123-155.
Van Mourik, J.M., 2001. Pollen and spores, preservation in ecological settings. In: Briggs, E.G., Crowther, P.R. (eds). Palaeobiology II. Blackwell Science, 315-318.
Van Mourik, J.M., 2007. Toepassing van optisch gestimuleerde luminescentiedateringen op enkdekken. In: Beenakker, J.J.M., Hortsen, F.H., De Kraker, A.M.J. en Renes, H. (editors); Landschap in ruimte en tijd. Askant, Amsterdam, 263-277.
Van Mourik, J.M., Dijkstra, E.F., 1995. Geen inheemse dennen rond de Oisterwijksche vennen; een palyno-ecologische studie. Nederlands Bosbouwtijdschrift 67-2, 51-59. (with english summary).
Van Mourik, J.M. van and F. Horsten, F., 1994. De Dijkerakker, een cultuurlandschappelijk monument. Weerter Jaarboek 1995, pp. 115-131.
Van Mourik, J.M. and & Horsten, F., 1995. De paleogeografie van de Valenakker. Weerter Jaarboek 1996, pp. 105-118.
Van Mourik, J.M and Ligtendag, W.A., 1988. De overstoven enk van Nabbegat. Geografisch Tijdschrift XXII-5, 412-420. (with english summary).
Van Mourik, J.M., Nierop, K.G.J., Vandenberghe, D.A.G., 2010. Radiocarbon and optically stimulated luminescence dating based chronology of a polycyclic driftsand sequence at Weerterbergen (SE Netherlands). Catena 80 (2010) 170–181.
Van Mourik, J.M. and Odé, B., 1990. Het Herperduin. Geografisch Tijdschrift XXIV-2, 160-167. (with english summary).
Van Mourik J.M. and Pet A., 2001. Broekbos of Ecobeek? Natuurontwikkeling in het dal van de Venloop. Nederlands Bosbouw Tijdschrift 2001-4, 12-16. (with english summary).
Van Mourik, J.M., Seijmonsbergen, A.C., Slotboom, R.T. and Wallinga, J, (2011a). The impact of human land use on soils and landforms in cultural landscapes on aeolian sandy substrates (Maashorst, SE Netherlands). Quaternary International 2011 / D-11-00088R2
Van Mourik, J.M., Slotboom, R.T., Wallinga, J., 2011b. Chronology of plaggic deposits; palynology, radiocarbon and optically stimulated luminescence dating of the Posteles (NE-Netherlands). Catena 84, 54–60.
Van Mourik, J.M., Wartenbergh, P.E., Mook, W.J. and Streurman, H.J., 1995. Radiocarbon dating of palaeosols in eolian sands. Mededelingen Rijks Geologische Dienst 52, 425-439.
Vera, H., 2011. 'dat men het goed van de ongeboornen niet mag verkoopen'; Gemene gronden in de Meierij van Den Bosch tussen hertog en hertgang 1000 - 2000.Uitgeverij BOXpress, Oisterwijk, netherlands. (with english summary)
Wintle, A.G., 2008 Luminescence dating: where it has been and where it is going. Boreas 37, 471-482.

6

Luminescence Dating as Comparative Data to Radiocarbon Age Estimation of Morasko Spherical Depressions

Wojciech T.J. Stankowski[1] and Andrzej Bluszcz[2]
[1]*Institute of Geology, Adam Mickiewicz University, Poznań,*
[2]*Institute of Physics, Silesian University of Technology, Gliwice*
Poland

1. Introduction

In the areas of glacial relief there are numerous small symmetrical endorheic depressions, the genesis of which is almost always described as cryogenic - melt-out, evorsive or post-periglacial. Occasionally different origin of these forms is described, such as karst or suffosive, while very rarely it is associated with impacts. Interpretation of the genesis of the latter is very difficult and requires specialized testing methods, among which the luminescence techniques seem to be very promising. They allow the authors not only to argument for the existence of an impact, i.e. the reset caused by temperature and pressure, but also the dating of the event.

The Morasko Meteorite Nature Reserve is one of very few places on the globe where, besides numerous lumps of metallic meteorites, the existence of impact craters was documented (Phot. 2, 3). There are serious arguments supporting this idea, although the genesis of these depressions is still controversial and some authors attribute it to cryogenic or impact origin (Hurnik 1976, Karczewski, 1976, Kuźmiński 1976, Stankowski 2009).

The elevations of Moraska Hill show glacitectonic structures which include Quaternary sediments of various age and the so-called Poznań clays from Neogene. Significant complexity of these structures means that the age of the rocks in the bottom of the Morasko depressions is from several million to ~18,000 years BP (Stankowski, 2009). This last date is a good reference point for radiocarbon dating of organic deposits as well as verifying luminescence dating of mineral fractions subjected to pressure and temperature of the falling meteorites.

The current state of knowledge regarding the age of glacial relief and the time of permafrost degradation in the analyzed area (Ołtuszewski 1957, Kozarski 1963, 1986, Makohonienko 1991, Tobolski 1991), clearly differs from the beginning of the sedentation in the Morasko depression (Tobolski 1976, Stankowski 2009) - see Fig. 1. This should be considered as one of the important indicators of the genesis of the controversial depressions which call into question their cryogenic origin.

2. Radiocarbon dating of the bottom layers of the organic deposits in the Morasko craters and peat sequences with metallic spherules in the melt-out depressions in the vicinity of Oborniki

Shallowly buried metallic meteorites, constantly found in the area of the Morasko Meteorite Nature Reserve and its vicinity, indicate their local fall. The results of the studies undertaken in the 1970s (Hurnik et al. 1976) show the north trajectory of the fall. Information about finding two small lumps of metal in the Obornickie Forests, which were lost during World War II, together with the results of the work of Hurnik and other authors (1976) formed the basis of the search for magnetic matter in peat filling the melt-out depressions and in organic and carbonate sediments that occur in the valleys of the Warta River tributaries (Fig. 2). Complementary data were obtained regarding the age of the peat sequence containing the admixture of fine-grained magnetic material as well as the age of the bottom layers of organic deposits which fill the Morasko depressions. This compliance is well documented by the results of the radiocarbon datings (see Fig. 1 and 2).

The initial series of datings was carried out at the Radiocarbon Laboratory of the Institute of Physics, Silesian University of Technology (A. Pazdur). The samples from individual depressions were of very diverse age. The earliest dates are 2.7 ka and 3.4 ka BP (see Fig. 1 – craters B and E). The samples were taken from the pilot drilling with light hand equipment in winter and posed the starting point for further specialized research.

Radiocarbon dating and the verifying luminescence analyses were performed on the drill cores from the two largest Morasko depressions – MOA with a diameter of about 90 m and MOB with a diameter of about 50 m (see Phot. 1). The cores were collected with the use of the equipment from GeoForshungZentrum from Potsdam thanks to the kindness of J. Negendank and the work conducted by M. Shwab. The radiocarbon datings were performed at the Poznań Radiocarbon Laboratory by T. Goslar - the calibration program OxCal v3.10 was employed (see Bronk 2001). The luminescence dating was realized in the Gliwice Luminescence dating Laboratory by A. Bluszcz.

The MOA profile, the depression permanently filled with water of ~1.5 to ~2.5 m deep.

- 0.000 – 0.250 m various peat from well preserved to fully decomposed with wood fragments and gyttja-like thin layers
- 0.250 – 0.290 m moss peat of diverse density; plant macrofossils fully recognizable; stratification visible through the section
- 14C dates - peat 20 cm above the bottom: 4,465±35 (Poz-18864); peat 3-5 cm above the bottom: 4,495±35(Poz-18863; 4,980-5,300 cal. BP)
- 0.290 – 0.295 m gyttja-like sediments

Very sharp boundary of Neogene/Quatrenary sediments in between peat and Neogene clays thin strata indicating almost immediate sedentation beginning after the origin of the depression

- 0.295 – 0.298 m silt with small quantity of sand
- 0.298 – deformed Neogene clays ("Poznań series")

The MOB profile, the depression permanently filled with water of ~0.4 to ~1.5 m deep.

- 0.000 - 0.385 m various peat from well preserved to fully decomposed with wood fragments and gyttja-like thin layers
- 0.385 - 0.390 m strongly decomposed peat/organic matter, slightly stratified
- 0.390 - 0.405 m decomposed peat with visible plant fragments and pieces of wood
- 0.405 - 0.407 m very dense, well decomposed peat containing some silty sand 14C date, peat 3-5 cm above the bottom: 4,760±40(Poz-18960; 5,320-5,600 cal. BP)

Very sharp boundary of Quaternary mineral and organic sediments

2-3 mm layer of dark greyish-brown fine sand.

- 0.407 - 0.447 m horizontally stratified silty, sandy, clayey sediments
- 0.447 - 0.472 m clay mixed with fine sand
- 0.472 - 0.672 m sand from fine to coarse with disperse organic matter
- 0.672 - Quaternary and Neogene sediments (clays, tills and gravels). The luminescence data obtained from the uppermost part of mineral sediments (~3-4 cm) beneath the organic matter - see results below.

Radiocarbon datings of organic sediments filling the Morasko depressions clearly indicate the young age of the early sedimentation - see dates 4,495±35 and 4,760±40 in above sections describe. Indirectly, this shows the time taken to generate the depressions, which is a few thousand years different from degradation of the permafrost and melting of buried dead ice masses in this part of the Wielkopolska/Great Poland Lowland Region (Kozarski 1963; compare also Fig. 1). This validates the view of the impact genesis of the analyzed forms - craters. The role of verification is seen in the luminescence analyses (Stankowski, 2011).

3. Luminescence data confirming a young, impact genesis of the Morasko depressions

Two approaches to luminescence analyses were used.

1. The first one was to determine the luminescence reset time due to the fall and plunge of hot lumps of meteorites into the sediments, which resulted in the creation of their sinter-weathering shells. The TL datings of four meteorite shells of the specimen weighing 10.5 kg, 11 kg, 21 kg and 164 kg, were realized by S. Fedorowicz from the Institute of Geography, University of Gdansk (Stankowski et al 2007, Stankowski 2009, 2011).

The dating results were similar, from 4.7 to 6.1 ka BP, corresponding with the previous palynological findings on the sedentation origins as well as the verifying radiometric data. The obtained TL values, indicating the reset time, document the local nature of the meteoritic shower in Morasko, and indirectly confirm the impact origin of the depressions, i.e. that they were generated in the form of craters.

2. The second issue was to find the reset scale of local Quaternary and Neogene sediments, which build the elevations of Morasko Hill. The above description of the geological profiles of the craters bottom layers indicate significant differences in the age of sediments - from the youngest, i.e. ~18,000 years BP and of glacial and fluvioglacial origin, to the advanced in age, i.e. the Neogene clays being at least several million years old. The age of the significant part of the deposits, and perhaps of the majority of them, should exceed the range of the luminescence dating.

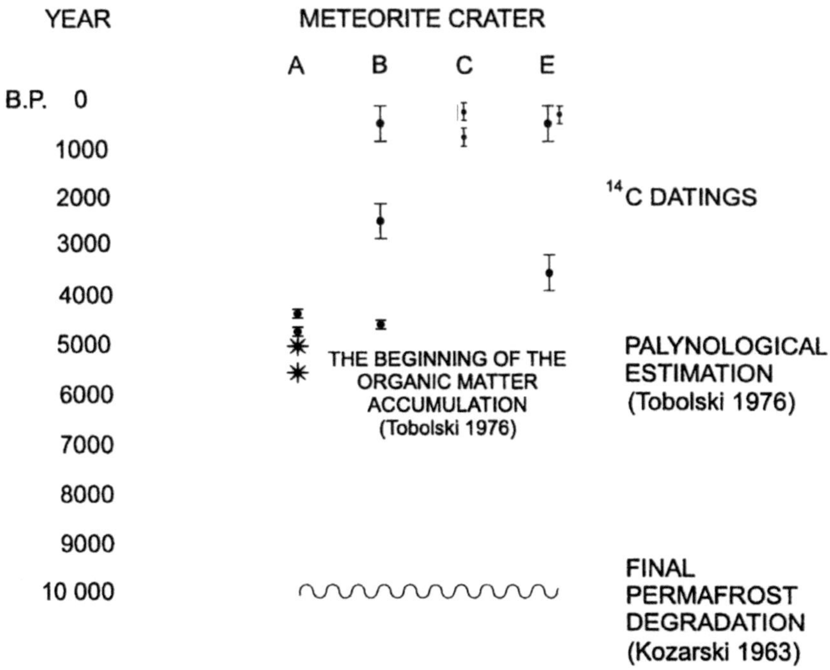

Fig. 1. Morasko depressions and their surroundings – radiometric dating, palynologically estimated beginning of sedentation, degradation of permafrost in the Wielkopolska/Great Poland Lowland Region.

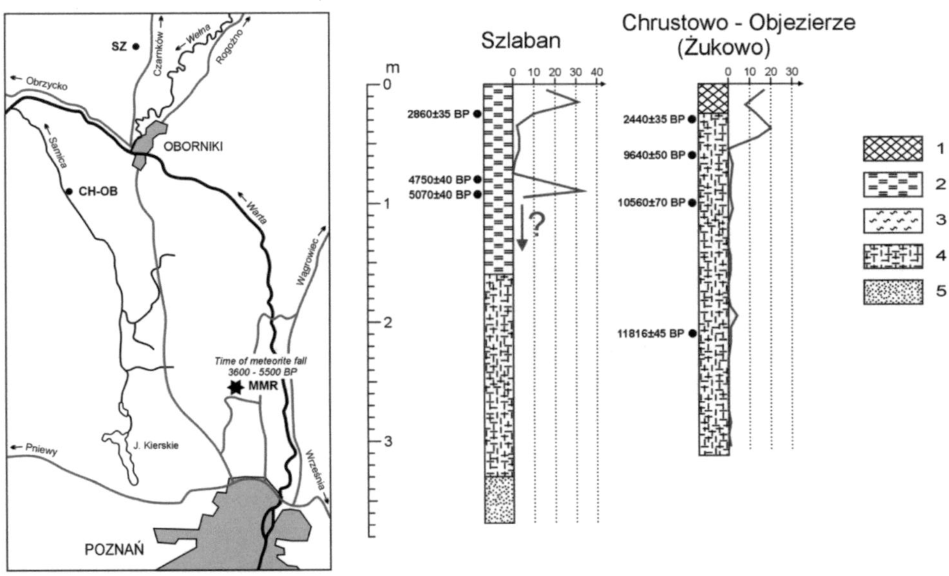

Fig. 2. Morasko Meteorite Nature Reserve (MMR) and its northern belt. Simplified geological profiles of the Szlaban and Chrustowo-Objezierze sites, with the 14C dates and indicators of content of small spherical magnetic fraction (spherules). 1) soil humus layer, 2) pea, 3) gyttja, 4) organic-rich silty muds, 5) sands with fine gravel.

The original unpublished results of luminescence dating in OSL technique obtained by A. Bluszcz (laboratory numbers for MOA crater GdTL-1328 through GdTL-1332 and for MOB crater GdTL-1333 through GdTL-1336), indicate the existence of a reset, but with varying extent. This is illustrated by the sets of luminescence measurements for the sediment samples from the bottoms of the both analyzed craters (Fig. 3, Fig. 4).

The results of the instrumental measurements in the bottom of the main craters shows the Fig. 3 and Fig. 4, with the containing tables.

Among the measured samples of sediments from the bottom of largest crater - MOA, build of the mineral Neogene deposits (Fig. 3), considerable variation of indicators can be observed. The collection of the oldest dates shows the range of the age from ~350,000 to ~45,000 years BP. A large number of indicators range from ~30,000 to ~10,000 years BP. Among the obtained data there are also many with the values lower than 10,000 years BP, with a significant proportion of them dated at <5,000 years BP. A statistically significant separateness of the ranges of measurement errors is worth mentioning. With respect to the young age indicators, the range of error turns out to be much lower. Thus, the mandated presumption is that, besides the nature of the measurement, it also stems from the effect of the reset range, potentially various in terms of pressure and temperature of a falling meteorite. Disproportionately low rates of the documented age of the sediments, in relation to their Neogene origin, prove very recent resetting. This rejuvenation, and especially the significant representation of the dates below 10,000 years BP, seems to justify the reset time of the deposits, and therefore the time when the Morasko Meteorite fell.

In the bottom of the second largest Morasko crater, MOB, Quaternary deposits are present. According to the authors these sediments are older than the last glacial period, at an age over ~130,000 years (the depression bottom is developed in sediments much older than the last glacial ones). The obtained age indicators of the 35 measured samples of deposits are much younger (Fig. 4). Only in particular cases they fall within the range from ~45,000. Other indicators did not exceed 27,000 years BP, with a significant proportion of the dates of <10,000 years BP. Occasionally, there are dates younger than 5,000 years BP. This range of dating seems to confirm the suggestion that the initial age of the studied sediments is older than the last glaciation, while the resetting time is younger.

The bottoms of the two analyzed craters are built of the sediments which are lithologically diverse and of very different age. According to the authors they originated more than 130,000 years BP. However, in relation to the earliest morphogenetic glacial processes, which potentially generate sediments, it is even possible to accept ~18,000 years BP, i.e. the time of the Leszno and Poznań phase of the last glaciation.

Taking into account the possible role of permafrost degradation and the time of the final melting of dead ice masses, we can theoretically go back to 11,000-10,000 years BP. Thus, the obtained indicators of luminescence age should only exceed this theoretical turning point. Meanwhile, both the Neogene clays and Quaternary sediments are often represented by very young luminescence ages. Approximately 47% of all the datings show indicators <10,000 years BP, among which ~19% are indicators of <5,000 years BP (compare Stankowski 2011). This indicates a very early time of luminescence resetting. The presence of older age indicators shows that resetting during the impact was characterized by a diverse range of completeness.

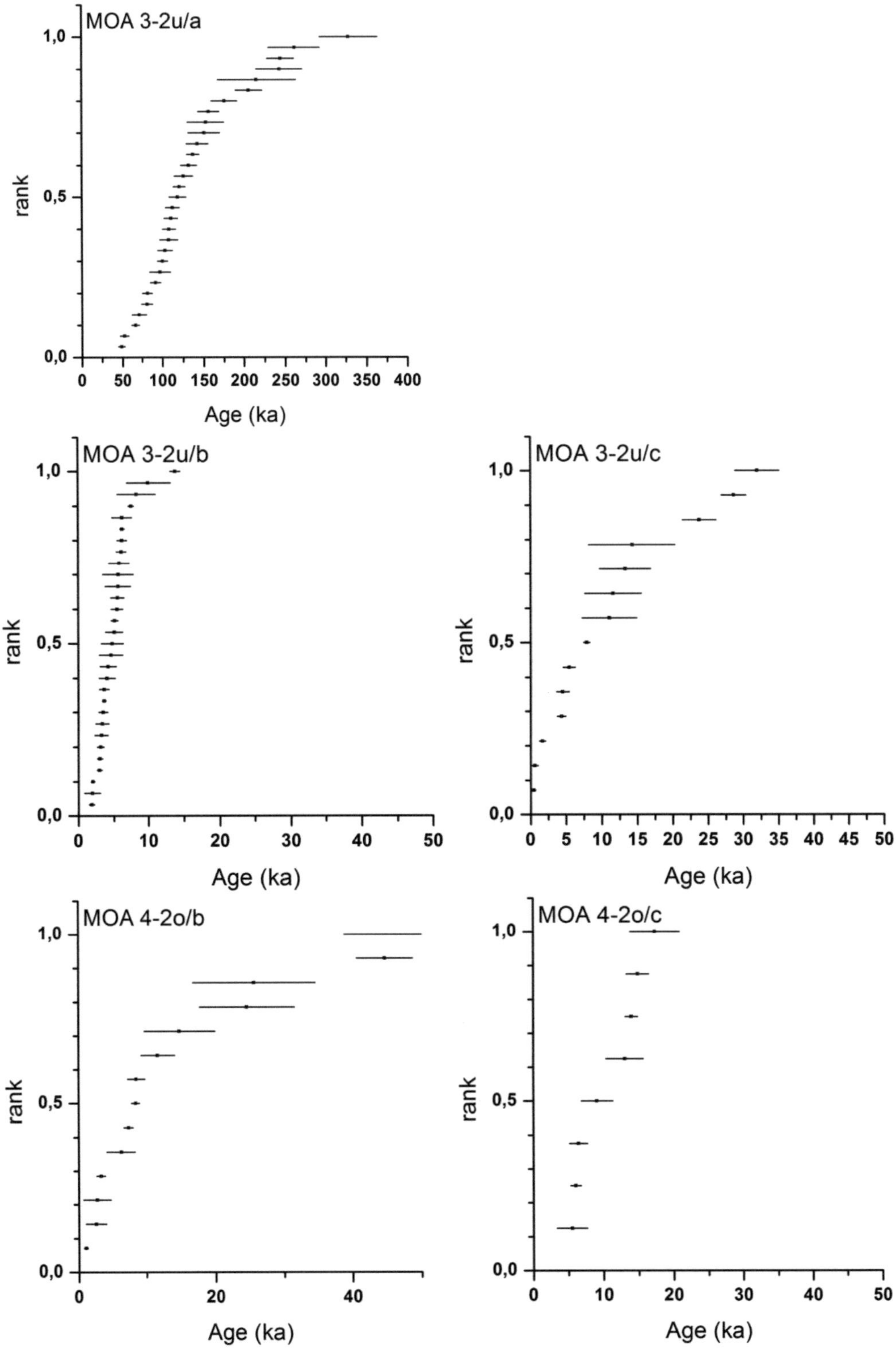
MOA 3-2u/a
rank
Age (ka)
MOA 3-2u/b
rank
Age (ka)
MOA 3-2u/c
rank
Age (ka)
MOA 4-2o/b
rank
Age (ka)
MOA 4-2o/c
rank
Age (ka)

Single aliquot OSL dating results (ka) for MOA samples

GdTL	MOA 3-2u/a 1328		MOA 3-2u/b 1329		MOA 3-2u/c 1330		MOA 4-2o/b 1331		MOA 4-2o/c 1332	
	Age	u(Age)	Age	u(Age)	Age	u(Age)	Age	u(Age)	Age	u(Age)
1	48,6	3,8	1,87	0,34	0,45	0,23	1,04	0,25	5,5	2,2
2	52,2	5,4	2,0	1,1	0,60	0,43	2,5	1,5	6,04	0,73
3	66,4	4,8	2,02	0,16	1,65	0,41	2,7	2,0	6,4	1,3
4	71,2	8,9	2,97	0,32	4,37	0,57	3,25	0,61	9,1	2,2
5	81,2	6,9	3,04	0,35	4,52	0,87	6,2	2,1	13,0	2,6
6	81,6	6,0	3,14	0,46	5,42	0,84	7,28	0,68	13,90	0,87
7	91,5	6,2	3,26	0,91	7,90	0,42	8,37	0,59	14,8	1,6
8	97	12	3,39	0,89	11,1	3,8	8,5	1,3	17,3	3,5
9	100,3	6,0	3,51	0,61	11,6	3,9	11,7	2,4		
10	103,5	8,7	3,66	0,16	13,3	3,6	14,8	5,1		
11	108	11	3,66	0,67	14,2	6,0	24,6	6,8		
12	108,1	8,0	4,1	1,1	23,7	2,4	25,6	8,9		
13	110,5	7,9	4,2	1,1	28,6	1,8	44,6	4,0		
14	112,4	8,2	4,6	1,6	31,9	3,1	51	12		
15	118,5	9,9	4,8	1,5						
16	120,7	7,0	5,1	1,2						
17	126	11	5,12	0,47						
18	132,1	9,6	5,49	0,81						
19	137,3	7,1	5,55	0,91						
20	142	13	5,6	1,8						
21	151	19	5,7	2,2						
22	153	22	5,8	1,4						
23	156	13	6,09	0,69						
24	176	16	6,17	0,66						
25	206	16	6,22	0,29						
26	215	48	6,2	1,4						
27	243	28	7,52	0,37						
28	244	16	8,3	2,7						
29	261	31	10,0	3,1						
30	328	35	13,82	0,67						

Other information

Sample	MOA 3-2u/a	MOA 3-2u/b	MOA 3-2u/c	MOA 4-2o/b	MOA 4-2o/c
Laboratory number	GdTL-1328	GdTL-1329	GdTL-1330	GdTL-1331	GdTL-1332
Number of aliquots dated	44	54	54	47	29
Acceptable dating results	30	30	14	14	8

Fig. 3. Indicators of the luminescence ages for the samples from the top layer of the Neogene clays in the bottom of the crater A (MOA). The horizontal scale - luminescence age indicator, vertical scale - cumulative frequency or rank. Some aliquots have not yielded a result when the absorbed dose could not be calculated from SAR measurements, and some calculated dose values have not been accepted when the assessed uncertainties were too high (high values of uncertainties are caused by weak OSL signals) or when an aliquot does not pass a recuperation test or a sensitivity correction test (Bluszcz, 2000).

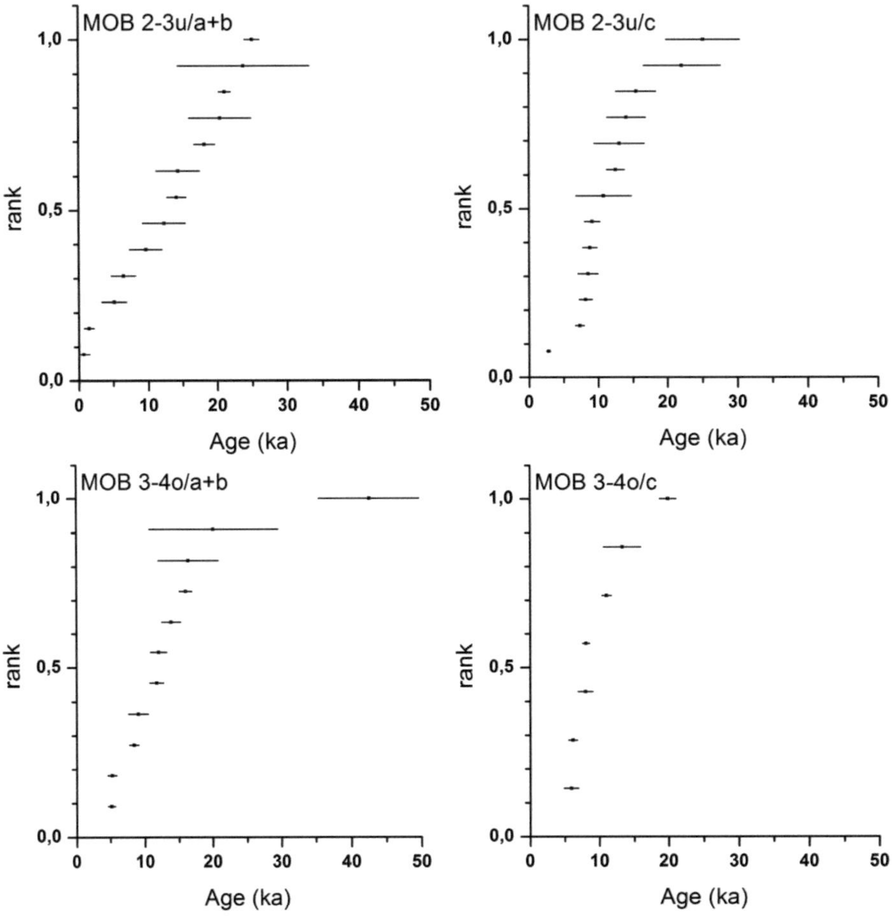

Single aliquot OSL dating results (ka) for MOB samples

GdTL	MOB 2-3u/a+b 1333		MOB 2-3u/c 1334		MOB 3-4o/a+b 1335		MOB 3-4o/c 1336	
	Age	u(Age)	Age	u(Age)	Age	u(Age)	Age	u(Age)
1	0,70	0,86	2,80	0,25	5,17	0,52	5,9	1,0
2	1,48	0,71	7,35	0,62	5,24	0,62	6,18	0,64
3	5,1	1,7	8,19	0,95	8,42	0,64	8,0	1,1
4	6,5	1,7	8,5	1,4	9,0	1,4	8,09	0,49
5	9,7	2,3	8,8	1,0	11,71	0,99	11,04	0,64
6	12,3	3,1	9,2	1,1	12,0	1,2	13,3	2,6
7	14,1	1,4	10,8	4,0	13,8	1,4	19,9	1,2
8	14,3	3,2	12,5	1,2	15,99	0,94		
9	18,2	1,5	13,1	3,6	16,4	4,4		
10	20,4	4,4	14,1	2,8	20,1	9,4		
11	21,04	0,84	15,5	2,9	42,5	7,1		
12	23,7	9,4	22,1	5,5				
13	25,0	1,0	25,1	5,2				

Other information

Sample	MOB 2-3u/a+b	MOB 2-3u/c	MOB 3-4o/a+b	MOB 3-4o/c
Laboratory number	GdTL-1333	GdTL-1334	GdTL-1335	GdTL-1336
Number of aliquots dated	44	40	39	34
Acceptable dating results	13	13	11	7

Fig. 4. Indicators of the luminescence ages for the samples from the top layer of the Quaternary deposits in the bottom of the crater B (MOB). The horizontal scale - luminescence age indicator, vertical scale - cumulative frequency or rank. Some aliquots have not yielded a result when the absorbed dose could not be calculated from SAR measurements, and some calculated dose values have not been accepted when the assessed uncertainties were too high (high values of uncertainties are caused by weak OSL signals) or when an aliquot does not pass a recuperation test or a sensitivity correction test (Bluszcz, 2000).

Photo 1. Morasko crater B (second in size) - obtaining cores from the bottom sediments. GeoForshungsZentrum equipment.

Photo 2. Fragment of meteorite of 7,5 kg found near crater C (third in size)

Photo 3. Meteorite of 164 kg found at the top part of southern fragment of the ringwall of crater A (main crater)

4. Summary

The results of the luminescence dating of the origin of the sinter-weathering shells, i.e. the resetting during the fall of the meteorite and its sinking into the sediments, were confirmed in the resetting of the material currently present in the bottoms of the depressions, i.e. the sediments were not moved from its original position and possible portions of the material gravitationally displaced from the slopes. This should be seen as a support of the concept of the impact origin of the analyzed forms. In the newly-created craters conditions for fast sedentation occurred. This is documented by the radiocarbon dating of the bottom layers of the organic deposits. Luminescence dating provided the verifying data of the genesis of the Morasko crates in relation to the previous radiocarbon and palynological estimations. The impact connected with the generation of craters took place in Mroasko ~5,000 years BP.

5. Acknowledgements

This study was possible thanks to the cooperation of many people, to whom we owe sincere thanks. We direct them to Prof. J. Negendank, PhD M. Schwab, from GeoForshungsZentrum in Potsdam, Prof. A. Pazdur from the Institute of Physics, Silesian University of Technology, Prof. T. Goslar from the Poznań Radiocarbon Laboratory, Prof. S. Fedorowicz from the Institute of Geography, University of Gdansk, as well as Prof. A. Raukasa from the University of Tallinn, for numerous constructive consultations.

6. References

Bluszcz A., 2000: Luminescence dating of Quaternary sediments - Theory, Limitations, Inetrpretations Problems (in Polish). Zeszyty Naukowe Politechniki Śląskiej, Geocgronometria 17 (seria Matematyka-Fizyka z. 86) 104 pp.

Bronk Ramsey C., 2001: Development of the radiocarbon calibration program OxCal. Radiocarbon, 43(2A), 355-363

Hurnik H., 1976: Meteorite „Morasko" and the region of the fall of the meteorite W: Meteorite Morasko and region of its fall. Wyd. Nauk UAM, Poznań 3-6

Hurnik H., Korpikiewicz H., Kuźmiński H., 1976: Distribution of the meteoritic and meteor dust In the region of the fall of the meteorite „Morasko". W: Meteorite Morasko and region of its fall. Wyd. Nauk UAM, Poznań 27-37

Karczewski A., 1076: Morphology and lithology of closen depression area located on the north slope of Morasko Hill near Poznań. W: Meteorite Morasko and region of its fall. Wyd. Nauk UAM, Poznań, 7-19

Karwowski Ł., Muszyński A.,: Multimineral Inclusions in the Morasko Coarse Octahedrite. 71st Meeting of the Meteoritical Society: Abstracts A71, 5232

Kozarski S., 1963: O późnoglacjalnym zaniku martwego lodu w Wielkopolsce (Late-Glacial Disappearance of Dead Ice In Western Great Poland - summary). Bad. Fizjogr. Nad Polską Zach., 11. PTPN, Poznań, 51-59

Kozarski S., 1986: Skale czasu a rytm zdarzeń geomorfologicznych vistulianu na Niżu Polskim (Timescales and the Rhythm of Vistulian Geomorphic Events In the Polish Lowland - summary). Czasom. Geogr., 57, Wrocław-Warszawa, 247-270

Kuźminski H., 1976: Dynamic element of the meteoritic shower "Morasko". W: Meteorite Morasko and region of its fall. Wyd. Nauk UAM, Poznań 45-63

Luecke W., Muszyński A., Berner Z., 2006: Trace element partitioning In ten Morasko meteorite from Poznań, Poland. Chemie der Erde 66 (2006), 315-318

Makohonienko M., 1991: Materiały do postglacjalnej historii roślinności okolic Lednicy (Beiträge zur postglazialen Vegetationsgeschichte In Lednica-Gebiet - zusammenfassung). Część II - Badania palinologiczne osadów Jeziora Lednickiego - rdzeń I/86 i Wal/87 (Palynologische Undersuchungen von Sedimenten des Lednicer Sees - Bohrherne I/86 und Wal/87 - zusammenfassung), W: Dotychczasowy stan badań palinologicznych i biostratygraficznych Lednickiego Parku Krajobrazowego. Biblioteka Studiów Lednickich, Wstęp do paleoekologii Lednickiego parku Krajobrazowego (red. K. Tobolski/ Introduction to Palaeoecology of the Lednica Landscape Park/Einführung in die Paläoökologie des Lednicer Landschaftsparks, Wyd. Nauk. UAM, Poznań, 63-86

Stankowski W.T.J., Raukas A., Bluszcz A., Fedorowicz St., 2007: Luminescencje dating of the Morasko (Poland), Kaali, Ilumetsa, Tsoorikmae (Estonia) meteorite craters. Geochronometria, 28, 25-29

Stankowski W., 2009: Meteoryt Morasko osobliwość obszary Poznania/ Morasko Meteorite, a curiosity of the Poznań region. Wyd. Nauk. UAM, Poznań, ss.92

Stankowski W.T.J., 2011: Luminescence and Radiocarbon Dating as Tools for the Recognition of Exstraterrestrial Impacts. Geochronometria 38, z. 1, 50-54

Tobolski K., 1976: Palynological investigations of bottom sediments In closed depressions. W: Meteorite Morasko and region of its fall. Wyd. Nauk UAM, Poznań, 21-26

Tobolski K., 1991(red): Dotychczasowy stan badań palinologicznych i biostratygraficznych Lednickiego Parku Krajobrazowego (Gegenwärtiger stand der paläobotanischen und biostratigraphischen Forschungen In Lednicer Landschaftspark - zusammenfassung). Biblioteka Studiów Lednickich, Wstęp do paleoekologii Lednickiego parku Krajobrazowego (red. K. Tobolski / Introduction to Palaeoecology of the Lednica Landscape Park/Einführung in die Paläoökologie des Lednicer Landschaftsparks, Wyd. Nauk. UAM, Poznań, 11-34